EYE OF THE STORM

Eye of the Storm

Facing climate and social chaos
with calm and courage

Terry LePage

Published by Open Door Communication
Opendoorcommunication.org

Cover design by Aleksandar N. Blowkowsky. Cover photo © Dmitri Vetsikas.
"Breathing Under Water" by Carol Bialock, from *Coral Castles,* © 2019. Used by permission.
"What Love Looks Like" by Terry Tempest Williams, *Orion,* https://orionmagazine.org/article/what-love-looks-like. Excerpt used by permission.

The publisher has made every effort to assure that website URLs referred to in this book are correct and active at the time of going to press. However, the publisher has no responsibility for the websites and can make no guarantee that a site will remain live or that the content will remain appropriate.

Every effort has been made to trace all copyright holders, but if any have been overlooked, the publisher will be pleased to include any necessary credits in any subsequent reprint or edition.

Cataloging-in-Publication Data

Names: LePage, Terry
Title: Eye of the storm : facing climate and social chaos with calm and courage / Terry LePage.
Description: Irvine, CA: Open Door Communication, 2023.
Identifiers: ISBN 979-8-9882257-1-3 (trade paperback) | ISBN 979-8-9882257-0-6 (ebook) | ISBN 979-8-9882257-2-0 (KDP paperback)
Subjects: BISAC: SCIENCE / Global Warming & Climate Change | SOCIAL SCIENCE / Activism & Social Justice | SELF-HELP / Stress Management

To my friends and fellow travelers

in the Deep Adaptation Forum

CONTENTS

PREFACE

In the past three years, I have supported hundreds of people who are facing or anticipating great loss from climate chaos, ecological devastation, and social upheaval. I have met them in my local community and in an international online group called the Deep Adaptation Forum, whose purpose is *embodying and enabling loving responses to our predicament*. I have witnessed their fear, their grief and despair, and their struggle to make sense of the strange new world unfolding around us. Species and habitats are disappearing before our eyes. Climate change, it turns out, means cataclysmic floods, epic fires, unprecedented heat waves and droughts, and other life-changing disruptions. Meanwhile, many governments and social systems seem to be crumbling before our eyes.

I have observed what helps people cope and what empowers them to take constructive action. I am inspired by so many people whose response to the breakdowns we face is to serve others, human and nonhuman. I participate in activist networks in my area but I don't expect to save much. Instead, I aim to help human and nonhuman communities live well, to proclaim the value and dignity of that which may be lost, or to act simply because it seems the right thing to do. I offer this book as a companion and guide to help you cope and respond to our predicament with compassion, creativity, and courage. Then you will find your particular ways of service and action.

I have worked as a research chemist, so I have some understanding of the scientific evidence around the climate crisis and the limitations of predictive models. I have been privileged to support people at the end of life and their families, and to train people in Nonviolent Communication as taught by Marshall Rosenberg. I bring my experience as a transitional minister in liberal Christian churches to this book. In times of transition, I have been the eye of the storm, the nonanxious presence who invites others into a space of calm reflection on how to put their values into practice despite an uncertain future. I know the power of framing our experiences through stories that express our values.

I wish for you, reader, to hold your ground amid fear and trouble, make the best choices you can with what you have, and make meaning and joy, whatever the state of the world. By doing this, you can be a refuge for others to shelter from chaos. You can be the eye of the storm.

Read this first: Is this book for you?

In 2018, I first began to take in the likelihood of severe and irreversible climate disruption in the next decade or two. At that time, almost all scientists and pundits were saying, "We can reverse climate change." Even now, researchers delivering the most dire predictions are expected to finish their horrific lists with an appeal to reverse the trend: "It's not too late." I find this stance dishonest. Although the time courses are uncertain, there is no more time to prevent massive losses of vast swathes of life on the planet, loss of global industrial consumer society, and loss of many human lives and cultures. I am watching the losses begin to unfold now. If you acknowledge this reality, this book is for you. If you do not know how to face this reality, this book may be for you. If you do not want to face this reality, this book is not for you.

Maybe you have already witnessed a beloved forest, farm, or river mined or laid waste for human gain, or seen a long-championed human or nonhuman rights cause fall to greed, scapegoating, or power plays. Maybe you grew up watching those in power over you extracting freedom and sustenance away from you and your community. If so, you have experienced one of the many kinds of collapses that have always been happening around the world. You already have valuable experience in knowing how to weather some form of collapse. I am sorry you have that hard-won experience. Know that you are not alone. The choking of ecosystems and the crushing of human rights and traditional cultures are often intertwined. Be warned: you may only be aware of a small part of what I believe is coming. Still, I hope you will join in conversations like the one in this book. Others may learn from your experience.

You may be contemplating possible futures, imagining the unimaginable, and it feels like looking over the edge of a cliff. That unmoored feeling comes in part from the loss of the stories we have told ourselves, stories about how the world works and our place in it. Now you see those stories for the lies they are, and you may be in existential free fall. Worldviews fall hard. You need a new story. Take your time to find one worth living by. Perhaps you will find a story or two to guide you in Chapter Three, Stories for courage in dark times. And please take time to study Chapter Four, Practical emotional support.

We do stand before a kind of cliff. It marks the end of what people like me in the industrialized world know and what we have relied on to live; things like somewhat predictable weather, birds in the trees, governments that more or less function, forested mountains, living reefs, the ocean off the streets, groceries on the shelf, and water in the tap. If you suspect that you will see the end of some of those things, you have glimpsed the cliff as we hurtle toward it. Recognize, though, that people are already making a life at the base of this cliff, and some have done so for generations. Where I live in California, some of those people are African Americans and Mexican Americans. The rest of us owe people who have lived this struggle deep respect, and we have much to learn from them.

If you see the cliff and feel overwhelmed, if you want to contribute but do not believe you can bear the emotional load, be kind to yourself. Do not force yourself to read more than you are ready to read. No "saving the world" is on offer here, so you are free of that burden.

If you feel that you are to blame for what is happening, this book may support you in shifting that view.[1] Industrial consumer society set up our Earth home for destruction, and those of us who grew up in this system are trapped in it; most of us do not know how to survive without it.

If you have had to flee your home to escape fire or flood, or were driven from home by violence or financial ruin, or have been living in fear or oppression because of your identity as a person of color, immigrant, or gender or sexual minority, you may already have stepped over the edge of that metaphorical cliff.[2] In this book, I'll go with you in spirit. I'll share a few stories like yours, mourn with you, and speak up for you. I can't bring you to safety. I can witness and affirm your worth, humanity, and creativity as you improvise to survive and live your values. I hope this book is of use to you. If you are already too familiar with the kinds of peril that await the rest of us, perhaps you can give this book, or portions of it, to the people in your life who do not yet understand your experience.

Maybe you refuse to acknowledge that any collapse is unfolding. I don't blame you. Why give up hope that sufficient numbers of world governments and their citizens will start acting responsibly, that limits and tipping points of our Earth home have not been exceeded, that technology will save us, and that life as we know it can continue despite the destruction of the global household that makes it possible? I don't blame you, but I can't join you. And this book is not for you.

—◦◦◦—

The ideas and tools in this book are first for inner adaptation, to aid you in adopting a worldview and some practices that will allow you to face disruption and hardship with calm and courage. With such a base, you can then choose the outer adaptation work to which you feel drawn or called. This book explores only a few of those choices and emphasizes one: with what you learn from your own experience facing unfolding crises, you can be the eye of the storm, the calm center, for others. We will need as much calm as we can muster to weather the coming storms with wisdom, compassion, and the least possible harm.

We are all in the same storm, but not all in the same boat. Throughout the book, I frequently use the term "we." From the context, it should be clear whether I am referring to all of humanity, to people who are facing our predicament, or to people who are living in or benefiting from industrial consumer society. I write from the perspective of a wealthy, university-educated, white woman in the U.S. with liberal views, and with some sense of the invisible knapsack of privileges I carry.[3]

You will find the perspectives of dozens of people in my local community and the Deep Adaptation Forum[4] international community. Most of these sections are excerpts of my interviews with them, or based on those interviews. Their voices provide a wider range of experience than I could give alone. I am deeply grateful to them for sharing their stories.

> Throughout the book, you will find quotes of voices other than the author's that look like this. This different style is used to aid the reader in recognizing long quotes and lists of quotes. Quotes may be from a person who was interviewed by the author or a comment on the Deep Adaptation private Facebook group. In that case, the words you see may be excerpted from the full quote. Brackets [like this] contain comments from the author.

This book touches on a wide variety of subjects without exploring them fully. Refer to the bibliography at the end of the book for more depth on some of those topics. If you are looking for a detailed guide to practical political action, community organizing for mutual aid, regenerative agriculture, ecosystem restoration, or other specific responses to our predicament, you will have to look elsewhere.

You will find a variety of writing styles in this book: essays, quotes, perspectives from many voices, poems, lists, and stories. If you are not interested in poetry, or technical stuff, or mental health tools, just skip those parts. You can read this book in order. You can skim or skip sections. Or you can consult the Table of Contents and choose a chapter, or even a section within a chapter, that appeals to you now, and a different part at another time. In this book, you will find dozens of ideas and suggestions. They are resources for you to choose or let pass. I hope some of them will support your own loving responses to our predicament.

- *Chapter one: Facing the storm*
  What we face, and the emotional impact of realizing what we face, in prose, poem, and story.

- *Chapter two: Stories shape us*
  Stories that don't work, and stories that may work, to frame our situation.

- *Chapter three: Stories for courage in dark times*
  Three stories to describe our predicament. They do not have happy endings. They have courageous endings.

- *Chapter four: Practical emotional support*
  Your toolbox for dealing with difficult emotions.

- *Chapter five: Befriending grief*
  Learning the skill of grieving to honor love and loss of all kinds.

- *Chapter six: Belonging and reverence*
  Finding meaning and wonder in relationships and in daily living.

- *Chapter seven: Resigning from the rat race*
  Slowing down and opting out of industrial consumer busyness and status.

- *Chapter eight: Connection and compassion*
  Tools for practicing compassion when it matters most.

- *Chapter nine: Letting go*
  Ways people are letting go of industrial consumer society and having fun doing it.

- *Chapter ten: No more flying solo*
  Ways people are building relationships for mutual support.

- *Chapter eleven: Young people and those who care about them*
  How we might be of service to those who inherit this mess.

- *Chapter twelve: Planting seeds*
  Gardening as activism and relationship.

- *Chapter thirteen: In the meantime*
  Ways people are living their values now.

- *Chapter fourteen: Endings are beginnings.*

Anxiety is contagious. Calm is contagious. And courage is contagious. Using the tools in this book can help you to find calm, purpose, and even joy in hard times. And you can share these things with others. You can be the eye of the storm.

Now, let us step into the storm together.

CHAPTER ONE

Facing the storm

This chapter introduces the predicament we face. It is more than a problem. Problems invite solutions. A predicament has been defined as "a difficult, perplexing, or trying situation from which there is no clear or easy way out." The chapter is made up of short sections on the following topics:

Sighting the storm. My dawning awareness of the depth of our predicament.

Not saving the world. The solace that comes from releasing ideas of fixing the unfixable. It sounds like giving up, but it's not.

Perspective: It's really that bad, and I am not alone. Kat's realization of the depth of our predicament and her discovery of a community that shares her awareness.

A healthy form of avoidance. A warning against listening to pundits who predict we will all be dead or nomadic shortly.

Perspective: Getting real. A list from Karen Perry of what we can still do after we envision a future of catastrophe.

Don't do this alone. My experience finding like-minded people and a plea to find yours.

Many voices: Living in two worlds. People share how they navigate living in the world of industrial consumer society's "business as usual" while holding an awareness of our predicament.

Breathing Under Water

I built my house by the sea.
Not on the sands, mind you;
not on the shifting sand.
And I built it of rock.
A strong house
by a strong sea.
And we got well acquainted, the sea and I.
Good neighbors.
Not that we spoke much.
We met in silences.
Respectful, keeping our distance,
but looking our thoughts across the fence of sand.
Always, the fence of sand our barrier,
always, the sand between.
And then one day,
– and I still don't know how it happened –
the sea came.
Without warning.
Without welcome, even
Not sudden and swift, but a shifting across the sand like wine,
less like the flow of water than the flow of blood.
Slow, but coming.
Slow, but flowing like an open wound.
And I thought of flight and I thought of drowning and I thought of death.
And while I thought the sea crept higher, till it reached my door.
And I knew, then, there was neither flight, nor death, nor drowning.
That when the sea comes calling, you stop being neighbors,
Well acquainted, friendly-at-a-distance neighbors,
And you give your house for a coral castle,
And you learn to breathe underwater.

– Carol Bialock[5]

Sighting the storm

I was in a cafe in San Luis Obispo, eating breakfast with my husband Scott, in the summer of 2018. We had driven on spectacular Highway 1 down the central coast of California, returning to Southern California from a visit to family in the San Francisco Bay Area. This two-lane highway winds precariously along the rocky shore on fragile sandstone cliffs. Highway 1 at Big Sur has been washed out repeatedly and was unusable recently for more than a year. Rockfalls and sections of one-lane road are routine.

While sipping our coffee, we were scrolling through news and social media feeds on our phones, as is our habit. I stumbled across a paper, self-published weeks before, by University of Cumbria Professor of Sustainability Jem Bendell, entitled "Deep Adaptation."[6] Over a tiny crumb-filled table, my world swayed. Puzzle pieces I had been collecting and filing away tumbled into place, forming a scene out of nightmare. Climate change was not some evil looming on the distant horizon. It was picking up speed, possibly about to upend civil society, and had already disrupted one thoughtful man's life. Strong coffee was not the cause of my shivers.

When I felt present in my body again, I was astonished that I had taken so long to realize what Bendell had written so bluntly. I had known about greenhouse gases since high school in 1977. I had been hoping Peak Oil, the decline of world petroleum production as reserves become harder to extract, would limit human carbon dioxide output to within livable ranges. I had foreseen the long game: at some point, I believed in the next couple hundred years, most of the ice on Greenland and possibly on Antarctica as well will melt or slip into the sea, some of it quickly. Industrial consumer society won't survive a roughly 215-foot sea level rise (66 meters). I fancied myself a far-seer. But it's one thing to see that on some far-off day, all our coastal cities and ports will be no more, and my home at 260 feet above sea level (80 meters) will be on a blocks-long island above an inundated city.

It's another thing to picture the destruction of cities in one's own lifetime. As Bendell's paper clearly spelled out, most of us have strong cognitive protections against acknowledging the likelihood of this reality. His paper removed those protections from me.

I returned to work as the transitional pastor of a small church. Those people loved to hear me proclaim the sacredness of Earth and our duty to care for her. They were not ready to hear about the end of the world. I never said that, exactly. I just kept going to dark places in spite of my attempts to moderate my preaching. Repeatedly, I cried in the pulpit. I felt the irony: for twenty years I had been preaching against Christian end-of-the-world thinking and the unsound theology of "rapture." (It's still unsound. God will not teleport anyone out of our predicament.)

If we were facing only the depletion of Earth's resources, such as oil, fertile soil, water, ocean fisheries, and pollinators, we might experience a slow erosion of industrial consumer society. We might have enough time and farm-friendly weather for sections of humanity to (re)learn a more Earth-respecting way of living. Instead, due to the baked-in heating of the planet, we are experiencing ever-increasing regional catastrophes across the globe from storms, fires, floods, droughts, crop failures, and heat waves. A barrage of local, regional, and specific collapses on an uncertain time frame against a background of more general decline seems to be in store, rather than one grand collapse. As these disasters keep multiplying, we will deplete the materials, supply chains, and goodwill we need to recover from them. People now experiencing local or regional collapses are already having trouble rebuilding something like their former way of life.[7] Social, political, and economic systems in many places will become dysfunctional. In some locations, they are already doing so; more is to come. This is the storm we face. I wrote the following poem in the summer of 2020, when the sky over my suburban California home was red with smoke from some of the four million acres of fires that burned in California that year.

You told me to just be.
You told me to pay attention.
You told me to be humble.
(I am not you.)

I didn't strive.
I witnessed.
What I saw broke my heart.
(A world lay inside one little heart.)

Then
I ran away.
I hid.
I put a tight lid on that unruly heart.
(But I can't live like that for long.)

Today
I walk gingerly
Shaking, crying,
Into the fire with you.
(The heart burns but is not consumed.)

Fear is contagious, calm is contagious, and courage is contagious. Those of us who have some idea of what is unfolding can prepare ourselves mentally, emotionally, and spiritually to be (as we are able) centers of calm, compassion, and courage. We can be ready to coach others to hold onto their values in hard times. Because we will have pre-processed some of the loss that others will deny for a while longer, we will be able to support them when they finally face what comes.

Not saving the world

Once I dropped from my shoulders the self-imposed burden of having to "save the world," I could breathe a sigh of relief and ask myself, "What can I still do?"

— Deb Ozarko

"Finally, I've found people who understand me." This response is typical in the groups I host. Hundreds of people have come to Grief Gratitude and Courage workshops, Deep Adaptation Welcome Circles, Compassionate Communication practice groups, Death Cafés, Grief Circles, Mutual Care Circles, and other groups I facilitate. Over and over, newcomers express profound relief and gratitude. They can voice their fears. They are not chided for "being a doomer" or "being too negative." They can be heard when they speak of the devastation they are witnessing, the despair and grief they are feeling. And they don't have to strive to save the world.

As I write, the term "collapse-aware" is in danger of becoming a tired meme. Yet it has served me well over the past few years. Collapse awareness means acknowledging that some things have gone too far to repair. The future, in one or ten or fifty years, will look radically different from the present, with losses that are hard to fathom. Saving the planet and happy endings are not on the horizon. There are no solutions for the complex predicament of interconnected unfolding tragedies we face. This sad truth is hard to swallow. Still, we can take constructive action.

When I first talked to my friends about my new awareness that climate chaos is irreversible and incompatible with industrial consumer society, they gently brushed me off. One touchingly honest friend said, "I can't believe what you are saying, Terry. If I did believe it, my world would be turned upside down." Indeed. Why subject yourself to that kind of thinking? Unless you want to base your actions on reality, not fantasy.

When I became collapse-aware in 2018, this was considered a fringe view. After the extreme fires, floods, and heat waves of the past few years, collapse awareness is no longer so fringe. But still, almost every worsening prediction by scientists, almost every dire news article, ends with an obligatory, "And here's what we must do to stop climate change." Those save-the-world lists allow authors and readers to bypass the reality of our predicament. They offer false hope, even as so many people begin to feel in their bones and their daily lives the unraveling of ecosystems and social systems.

If we accept that huge losses in the human and nonhuman world are inevitable, then what? Are we giving up on humanity or on the planet? No. But it will look to those around us as if we are giving up. We may feel like we are giving up because so many things on those save-the-world lists no longer make sense to do. We are giving up on wishful thinking. The goal is no longer to save the world. The goal is to love the world and the humans and nonhumans in it and to care for them as best we can, while we can. Each of us will express that love and care in different ways. Some ways will be practical, some spiritual. Some ways will be political, some entirely earthy and home-bound. Some ways will be very local, and some will be wider in scope. Our actions can affirm the value of human and nonhuman life, ease suffering, and foster supportive community instead of cruelty.

This shift from the goal of saving the world is hard mental work because it requires letting go of the stories of modernity that are embedded in us, those of us raised in industrial consumer society anyway. More on that in the next chapter. Giving up on saving the world is also hard emotional work. Coming to terms with the immense losses it implies is heartbreaking. Further, it is hard on the ego: do I have to let go of so much? Can I really only do so little? The industrial consumer ego has been vastly over-inflated.

As they decide how to live honorably and well with the reality of loss, people are discovering new priorities and new values. Better understanding our limits means that the loss of a job, a personal project, an organization, a home, a beloved landscape, a human right or freedom, or a community, may be a part of the process of collapse.

As hard as it is to live through, it is not a personal failure or a sign of incompetence. Knowing that any of us can experience such a loss, we can commit to supporting each other through hard times. And we can learn to live as if Earth matters, with humility and gratitude, as if we are one small part of a wondrous planetary metabolism, our Earth home.

Perspective: It's really that bad, and I am not alone

Kat has been the Coordinator of the Deep Adaptation Forum. She hosts and participates in as many gatherings in the Forum as she is able. She also runs a nonprofit ecological consultancy and makes a home with her family on an acre in the north of Scotland. She is sometimes found in her new garden, scheming, digging, chasing chickens, or weeping. She is often found in one of the Forum's many Zoom meetings, holding space for people facing our predicament. These are her words about discovering others who thought like her about the state of the world:

> Sustainability, the climate crisis, and ecological challenges of biodiversity loss and extinctions, that's been my career. Because of the circles I move in, Jem's paper "Deep Adaptation" came to my attention within just a couple of weeks of it being published in August 2018. I was in the car park at my office, about to start my drive home. An email came in with Jem's article. I clicked on it. I thought, "I'll just have a quick scan and see if it really is something I want to read. And then I'll put the music on, and I'll make my 90-minute journey home, and I'll have dinner with my husband, and everything will be lovely, and I can forget all the troubles of the day." I opened the article to start scanning it, and I was still in the car park *three hours later*. And then I sat still for a long time. I just thought, "Someone said it out loud." Finally, someone, a voice from within the system, has written it down and said it out loud. I was overwhelmed.

I wrote to Jem on that day. I sent an email saying, "I can't believe what you've done. It's paradigm-shifting for me. Congratulations on having the courage to put yourself out there to actually write all this stuff down and say it in public." Because these aren't things that you were allowed to talk about within the sustainability sector. It was unacceptable. You know, quiet, hushed conversations at the back of a conference room. Questioning or challenging some assumption that had been made was the closest you could get. For my entire career, I've held the knowledge that the system was probably going to collapse within my lifetime. I'd held that on my own. Me, thinking, "Maybe I'm mad, what do I know? I'm just a scientist. I'm not a global expert on anything." So there was always that second-guessing of the self. But you feel it. Sensitive people, we might not ever be able to articulate it or describe it. But there's a sense of the *dis*-ease in your system as you observe the world around you. It can be in really simple situations. There's a dis-ease in you. This is not right. This is not right. Jem's paper was saying, "Damn straight, it's not right. Here's a really good illustration of all the ways it's not right and where they're going."

So when I discovered that the Deep Adaptation Forum was being convened, I joined. Within a day or two, I'd had an email welcoming me to the community and asking me, Would I be prepared to volunteer? And here are some of the gaps that they were looking to fill... So I agreed to volunteer. I felt like I had found my tribe. I was 49, and suddenly it was okay to talk about the things that I'd held in my body since I was 13 years old. Not only did I now have a vocabulary, but I also had a community of people with whom I could share this knowledge. It was joyful, and it was hideous. It was joyful, because of the ability to make those connections, because of that sense of not being alone. And it was hideous, because although I'd carried that knowledge in my system throughout all of my life, I hadn't really accepted it, because I hadn't heard it from anyone else. No one ever spoke about that dis-ease. No one ever spoke about their observations about how the world was unraveling.

So it was a mixed blessing to find the Deep Adaptation community. There was this relief. "Ahh, it's a bunch of people who feel the way I do." And then there was the, "F**k, there's a bunch of people that feel the way I do. This is f**king real." The grieving process, particularly in the first year, was at times debilitating. I remember being in the car and seeing something and thinking, "For how many years will I observe that?" This poignant, beautiful moment, and having to pull over at the side of the road and weep wracking sobs of grief over what is being lost.

The comfort, camaraderie, fellowship, and belonging have strengthened in the three years that I've been part of the community. And the grieving has never eased. Now I grieve more, deeper, and for more things. I have found [my participation in the Deep Adaptation Forum] to be strengthening. It's enhanced and built my capacity for being with what's broken and messy and untidy. My capacity to be with grief has grown sufficiently so that it's never overwhelming. And it's inspired me to do things with my community for resilience as well.

A healthy form of avoidance

Do you feel obligated to attend to all the tragedies of the world? You are not. Truly. Nor are you required to be up on predictions of future catastrophes. Know how much exposure to suffering that is not your own is too much for you to process. Then do not subject yourself to that much.

Thanks to the internet, negative news is relentless. It is easy to fill our eyes and ears with stories we don't need to know. We were not meant to hold the disasters and outrages of the whole world in our tribal-sized minds. Notice which news sources and types of articles are informing you of something new versus articles that are redundant or mostly gossip. Consider limiting disturbing news and social media posts to a small portion of your day, or even of your week, so that you have time to live, love, and create.

Occasionally I run into people who are sure that humans will be extinct in five to ten years. They have usually been listening to Guy McPherson and his colleagues. Things are bad and getting worse, true. The Intergovernmental Panel on Climate Change (IPCC) underestimates the severity of our predicament every time, true.[8] Yet we do not know enough to predict a schedule for our various collapses.[9]

Seeking a clearer understanding of our predicament, I listened to a video presentation by Guy McPherson. I had hoped to gather data and graphics showing the science of climate and ecological collapses, present and predicted, to include in this book. I found none of that. All I got was the punch lines from various scientists and pundits. "This system will collapse." And, "That system will die." His report contained no science, nothing informative, just pronouncements of doom. The worst prognosis from each scientific and nonscientific report was clipped out of context and read by McPherson. He has already predicted several demises that have failed to come on schedule. Among these, in 2012, he predicted that global warming would kill much of humanity by 2020.[10]

While our minds crave certainty and our predicament offers little, creating false certainty is not helpful. Therefore, I avoid Guy McPherson and his term "Near Term Human Extinction." Our Earth system is very complex and defies accurate predictions. Each of us faces personal extinction at our death; we always have. Loss of a way of life, a life, or even many lives is far from the extinction of the human species. Let's live, grow, create, and love in the face of uncertainty.

Another author, Michael Dowd, finds it helpful to accept the possibility of worst-case scenarios. Yet he does not spend his time making predictions. Instead, he ponders ways to make sense of collapse, ways to cope, and ways to be of service. He writes, speaks, curates, and records the work of others at the website postdoom.com.

Perspective: Getting real

Karen Perry stewards the Chickenfoot Ranch in Northern California with her partner Jordan, and expects the worst. She wrote the following guide for responding to our predicament. She calls it GRAC/E: Getting Real about Collapse/Extinction. She writes:

The civilizational way of living has never been sustainable. The 10,000-year experiment of living differently, separate from the rest of the community of life, has always risen and then fallen. Global industrial civilization is collapsing, just as every civilization model before this one has done. This time, however, the harnessing of fossil fuel energy has made our species "Homo colossus," giants with the ability to destroy the natural systems required for our survival. The result is, in addition to societal collapse, the biosphere we depend on for life is also collapsing. This is a predicament, not a problem, thus requiring wise responses, not false solutions.

1. Before the truth will set you free, it will likely make you angry. Don't shoot the messenger.

2. Let go or be dragged. Get through the grief and know that freedom and benefits exist in acceptance.

3. Grab a buddy who won't pull you down. It's harder to get real alone. Expect other crabs in the pot to constantly yank on you. Resist the urge to pull others down if you slip back into denial.

4. Abandon Hopium. Hope and false solutions are placeholders for inaction. Both are harmful responses to our predicament and can foster even more denialism.

5. We all would benefit from becoming much more comfortable with death and dying.

6. Nature is primary. Period. No more humans first. Making amends to the rest of the community of life, while attempting

to clean up the mess, needs to be a daily life way. Every day is Earth Day.

7. There needs to be an urgent conversation about setting the younger generations free and what that can look like. Continuing to pass along the dominant culture denialism is super abusive.

8. If the response feels predictable and familiar, it's not the right response. Our species is really being challenged to elevate. It's not about becoming more clever. Predicaments do not have solutions. We need radical responses equal to this radical situation. Logic says if it's a cancerous growth and growth is the issue, the first response is to stop growth.

9. Apply GRAC/E (Getting Real About Collapse/Extinction) [this list] to daily living and let it guide all decision-making. If you have privilege left in the game, what will you use it for?

10. Bottom line, we have to get comfortable talking about it. Continuing to bury our heads in the sand is not the path to enlightenment. People say, "Well if this is happening then there is nothing we can do about it so why think about it, too depressing." I say see numbers 6 [*nature is primary*] and 7 [*setting the younger generations free*].

Don't do this alone

I believe that the community—in the fullest sense: a place and all its creatures—is the smallest unit of health and that to speak of the health of an isolated individual is a contradiction in terms.

– Wendell Berry

Find your people. Find the people who will support you through hard times, and through the changes you want to make to face hard times. And recognize that you are a small part of a vast interconnected Earth home that is sustaining you. You will hear this call to human and nonhuman community many times in this book because it is so essential.

Industrial consumer society relies on the illusion of separation, the belief that people are independent of the human and nonhuman systems that sustain us through long and often exploitative supply chains. We in the U.S. have made independence a virtue. And no wonder, in a society where most of our interactions are transactional, buying and selling, owing and paying debts, and competition for status and stuff, rather than respectful and reciprocal relationships. We have paid a dear price: a cultural epidemic of loneliness, and the loss of many of the skills and strategies of our ancestors that allowed them to live and work together for the common good.

To face the storm that so many ignore, we need people who understand us. Few people can sustain a new identity and values in the midst of the old unless they spend time with people who support those new values and that new identity. They need not be our family, neighbors, or closest friends. Who supports you being you? Who can you talk to about how you might want to live, about what matters, and how you want to show up in these times? Who is already living the values you want to live?

Respectful relationships with nonhuman entities have been devalued, suppressed, or forgotten by industrialization, colonialism, and extractive capitalism for generations. That forgetting has allowed industrial consumer society to ravage the planet. The barriers to recovering these essential relationships are formidable. Modernity does not even allow language for the interconnection of the human and the nonhuman. Attending to relationships with the nonhuman world with respect and wisdom is new to me. That attention can look many different ways, but it is a key practice for affirming that we all rely on our Earth home for belonging and sustenance.

For the past two years, I have been talking with people around the globe who are exploring ways of living outside (or in the cracks of) industrial consumer society. The first gift of these online friends was to support me in grieving our losses, even as I supported them, and to plot together ways to live and love in the face of loss. This book is possible only because of my friends in the Deep Adaptation Forum. These friends have changed me in other ways. Some of them refuse to fly. I had been in the middle of the pack with my (mostly well-off) friends. They profess concern about carbon emissions but fly for business or pleasure every chance they get. Now I think long and hard before flying.

Modern life makes it easier technologically to create groups of like-minded people supporting each other. And it leaves us without the relational skills to sustain these groups. Cultivating and maintaining relationships requires time, skill, and commitment. Industrial consumer society has convinced us that time is for producing and consuming, not for building relationships. Still, finding your people is necessary to begin to live outside the values of industrial consumer society.

We will return again and again to this challenge of community and solidarity in the eye of the storm.

Many voices: Living in two worlds

Once we know about unfolding collapses, we can't un-know them. And we probably can't force others to know them. Facing the denial in our culture, we will have to figure out how to live in two worlds. If you are struggling with this, you are in good company.

Someone asked on the Deep Adaptation private Facebook page, *"How do you psychologically manage doing "business as usual" (driving in your gas-burning car to your capitalism-serving job...) and being collapse-aware, wanting to or trying to live as if earth mattered?"* People from around the globe answered. Here are some of their responses:

- I don't. I just have frequent breakdowns.

- I call it, "living in the twilight zone." I find it difficult every day.

- This conundrum has plagued me for years. My head tells me, "Oh s**t we need to do something," and my actions seem to say,

"Another day, another dollar." It causes me to feel self-loathing at times.

- It is a surreal feeling, to say the least.

Here is how some people navigate the cognitive dissonance of living in two worlds.

- It's called compartmentalization. I know it [the reality of collapse] is there all the time, but I need to distract myself from it in order to live life. It's hard.
- For me it's distinguishing my work from my side hustle. My side hustle is "the business as usual" that I need to do to stay alive. My work is to heal my inner realm, to transform my trauma, leading me to find my life purpose.
- Managing small steps I can tackle on my own that contribute gives me some calm. I live in Southwest Florida. Need I say more? The only thing I can control is my reaction to anything.
- Take some action. Even a very small action will make you feel more positive. Even just having a plan to take action is a start.

People offered particular actions that help them live in two worlds.

- I wish I could stop participating, but then I don't know how I would support my family. It's not like we can be hunter-gatherers anymore when every acre of land is owned by someone now. So I am trying to build an eco-friendly business and helping to start a community garden.
- Sending emails to politicians. To companies. Showing up for every town meeting I possibly can. Showing up to every political meeting I can. Organizing.
- I work in solar. So even though I know collapse is coming, I know that my work helps buy us time as a planet. It helps people save money, and get better prepared for emergencies.
- I walk most places. It's certainly limited my "reach." I look at it as a psychological adaptation to collapse. I feel like a good example of living differently, my version of "in this world, not of it."
- I do hospice volunteer work.

Some people have chosen to quit their jobs, or live very simply, in response to the tension of living in two worlds.

- I quit that stupid job in 2020 and got a permanent work-from-home job. It's still 40 hours a week, which is unhealthy, but at least I don't waste any time commuting. I have more time to try and grow food and do other household stuff that keeps our consumption to a minimum.

- For me it was important to stop. I didn't have dependents so I could handle financial and housing vulnerability. I closed my company (didn't sell it as that would have meant its ecological impact continued.) I moved into a low-impact tiny house, swapping a little labor each week for use of land. I use my time growing vegetables and offering support and information to others on how to navigate these times, how to need less money, and adaptation.

Living in two worlds, we face deep loss, both ongoing and anticipated, and a whole culture surrounding us that refuses to acknowledge that loss. That is why grief work is a necessary part of being the eye of the storm. We need this work; grief is addressed later in this book.

Strangely, the loss that may most disorient people is a loss of meaning. Cultural stories that once defined and guided us no longer ring true. So part of being the eye of the storm is discovering different stories that are not part of the systems of denial and destruction, but instead are life-giving in this time, stories that affirm meaning and identity despite great loss. The next two chapters are about stories, the stories that are killing us, and some stories that invite us to live differently.

Summary and reflection

- In 2018 I came to a gut-level realization of our unfolding predicament. That realization came with an awareness that "saving the world" as we know it does not seem possible.

- In groups I have hosted, participants express relief that they don't have to pretend that we can save the world as we know it. Together we are finding meaningful ways to live and love in the face of great loss.

- Kat knew from adolescence that business as usual was not sustainable. She remembers the power of discovering that she was not alone in her thinking. She now has a community to support her in this heartbreaking reality.

- We don't know enough to make accurate predictions about timelines for the losses I am calling collapse. As uncomfortable as it is, living with uncertainty is better than making dire predictions that send people digging their own graves.

- Karen Perry's guide to GRAC/E, getting real about collapse/extinction, in ten steps, is practical and wise. There is much good work to be done.

- Strong community, human and nonhuman, is essential to meet the unfolding storms. Fear is contagious, calm is contagious, and courage is contagious.

- Find the people who will support you through hard times and through the changes you want to make.

- Living in two worlds, that of business as usual and that of collapse awareness, is not easy. People are using various strategies to cope and to live differently.

CHAPTER TWO

Stories shape us

In this chapter, we will examine cultural stories: stories that serve life and those that don't. We will challenge stories you may take for granted and explore stories that honor interdependence with people and the nonhuman world.

The power of stories. Telling stories is the best way to convey meaning.

Stories of disconnection, stories of interdependence. A list of stories from industrial consumer society that no longer serve us, and suggested replacements.

Perspective: Pop culture stories. At fan conventions, Daniel shares his reflections on popular fiction that he finds helpful for understanding our predicament.

Hidden stories. Approved histories have left out the parts where nature and powerless people were destroyed.

Nihilism as a failure of stories. How to know when your stories are failing you.

The limits of science. Some reasons science has failed to address our predicament.

Perspective: One shattering detail. What the International Panel on Climate Change didn't tell us changed Tim DeChristopher's life.

Perspective: The story nobody wants to hear. Sven found holes in biofuels, climate predictions, and tech fixes to our predicament.

Perspective: Stories to PLAN by. Academics can tell stories, too; here are some good ones.

Choose your stories wisely. Alternative stories require careful examination. They may have issues of their own.

The power of stories

We tell ourselves stories in order to live.

– Joan Didion

When asked a question, modern people are likely to give a linear, factual answer. When asked a question, elders in a traditional culture are more likely to answer with a story. The story does not answer the question but rather shows a valued principle by which the questioner can be guided. Such stories feel quaint to moderns, but they can take root in us in a way that linear answers seldom do. Stories are an ancient and effective way of knowing. I turn to stories to explore the ways of thinking that led us to this mess, and the ways of thinking that can redirect us toward more life-giving practices.

Some stories appear to be untrue because they are somebody else's foundational myths. Some stories appear not to be stories at all, but axioms about how the world works. These "axiomatic" stories are *our* foundational myths.

My culture's foundational stories shape and direct us, marking what is valued, and even what we think is possible. Whatever their benefits, these stories have led to the destruction of our Earth home. The values in these stories do not work in some fundamental and definitive way. Letting these stories go will not be easy. They are burned into our brains, and we act unthinkingly according to them. But we can begin to unlearn.

Stories of disconnection, stories of interdependence

There are these two young fish swimming along and they happen to meet an older fish swimming the other way, who nods at them and says, "Morning, boys. How's the water?" And the two young fish swim on for a bit, and then eventually one of them looks over at the other and goes, "What the hell is water?"

– David Foster Wallace[11]

Certain stories have helped create our predicament. At root, they are all *stories of disconnection*. They describe people as isolated individuals, neither reliant upon nor responsible for our Earth home and the humans and nonhumans that are a part of it. Denying our interdependence with human and nonhuman lives and our Earth home has allowed industrial consumer society to destroy them. In this chapter, I present a few of these stories in the form of statements and principles. This bare form may not show the allure of these stories. Or maybe they never were alluring, but simply part of the air we breathe. Each story has a little commentary and a proposed alternative, also in bare form.

The alternative stories are all *stories of interdependence*. You could also call them *stories of kinship, stories of belonging, stories of mutuality, stories of interbeing,* or *stories of reciprocity*. I choose *interdependence* not for its poetry; sadly, it has none. I would like to choose one of these other words, but my understanding of their full meaning is impoverished because of my disconnected culture. *Interdependence* makes it clear that we are utterly dependent on each other and the nonhuman world for our lives, a reality that my culture desperately needs to learn.

Some stories of disconnection are spoken aloud, and some of them act in us, below the level of conscious thought. The assumptions of modernity are fundamental to our identity. Like the fish who don't know water, we do not recognize these stories as the basis of our living, let alone as one set of stories among many possible framing stories for making a culture and a life.

When our stories fail to explain our experience, we become disoriented. This is why most people can deny the facts in front of their noses so easily. *It is easier to deny what one is seeing than to deny the stories that define one's reality.*

Which stories are failing you? Which stories did your culture teach you that gave you meaning and purpose for a while, but now seem pointless in light of our trajectory? Maybe you are starting to realize, or have known for a while, that a story you once took as truth is no longer one you want to live by. Maybe you just feel soul-sick, and you don't realize that you are trying to live by a story that is no longer life-giving, or never was. Which stories prevent you from showing up with love and creativity in this strange and fearful time? A story that is life-affirming in one circumstance, or at one time in your life, may stimulate despair in another.

I do not expect the alternative *stories of interdependence* to transform our death-dealing systems. My aim is more modest. I want to be freed from living mindlessly in stories of disconnection. I want to envision other ways of being, and perhaps practice some of them on whatever scale I can manage. I invite you to join me in these goals.

Whether these stories are true or false is not the question I want to ask. *Each story invites a way of being in the world.* Each is true from within the world it helps create, and false when viewed from a different way of being. Is a story life-giving for us now? Sit with each *story of disconnection* and its suggested replacement *story of interdependence*. If the story of disconnection sounds familiar, give it flesh by thinking of situations in your experience that illustrate that story. Take your time.

——⟅ᴧᴧ⟆——

Story of disconnection: Get a good education, work hard, and you will have a secure future. Financial success is a reflection of worthiness and work ethic. And implied: financial failure is a result of worthlessness, laziness, and/or incompetence.

My parents taught me this story. It worked for them, and it worked for me and my husband. For many young people, it no longer works. For many other people in the U.S. and around the globe, financial security was never possible. My financial security has required their low-wage jobs and expendability.

Teens and young adults from privileged backgrounds are struggling frantically to try to fulfill this story. More and more of them are suffering from debilitating anxiety or depression. It's almost impossible to start a "middle-class life," so many young adults are bitter or despairing, and labeled failures.

Story of interdependence: We live in uncertain times. There is no route to a secure future. But your work is honorable if you are creating something, maintaining peoples' well-being, or serving the nonhuman world. You will be less burdened if you live more simply than your parents. No one can guarantee your security, but you can find meaning and satisfaction through creativity, relationship, integrity, and service. We treasure you, and we need your contribution, small though it may seem to you.

This cultural message is desperately needed. If you don't use any of the other stories, please take this one to heart and share it with the young people you know. Share it with people who face financial ruin, loss of job or housing, or people who are buckling under the stress of trying to hold onto a job that pays the bills but crushes them, body and spirit. It is honorable to be poor in an economic system that is eating itself.

—◈—

Story of disconnection: Competition is good. It brings forth the best in people. Furthermore, complex societies are inevitably hierarchical; that is the price of civilization. The ancient story that underlies this, which democracy denies but capitalism relies on, is this: a hierarchy of value exists for all beings. People are at the top. The rich and the rulers are at the top of the top.

Cultures can teach and value competition, or cooperation, or a mix of both. When competition is emphasized, selfishness is rewarded. People don't acknowledge or respect the larger web of human and nonhuman entities that sustain them.

Capitalism is a particularly brutal form of competition. It makes acquiring money and possessions the highest value. Those who accumulate the most are declared heroes, even if they are destroying lives and livelihoods in the process with ever more extreme schemes of destruction that affect us all. Unrestrained capitalism is sociopathy, and it is celebrated in my sick culture.

Story of interdependence: Exploitation and hierarchy are not inevitable parts of complex cultures. Cultures that respect and care for humans and for the nonhuman world must be, to a great extent, cooperative.

Moderns have a lot to learn about cooperation. David Graeber and David Wengrow have shown that many large, complex ancient cultures have not been as hierarchical as researchers assumed.[12]

Cooperation is usually the practice of communities that recognize their interconnectedness and reliance on one another. Such communities are often labeled "poor" by researchers because they do not hoard, and they exert social pressure on anyone who takes more than their share. They do not require the existence of a losing lower class beneath them to support their winning. They prioritize relationships. By practicing mutual aid, they are often able to insulate each other from suffering caused by a personal loss or lack of sustenance. Many remaining traditional cultures fit this description, as do some marginalized communities within industrialized settings. Mutual aid, and all work that does not get paid, do not show up in measurements of wages or GDP. So perhaps these people are not as poor as they appear.

Story of disconnection: Growth is good and necessary. We need jobs. We need healthy retirement accounts. More is better. More stuff, more status, more wealth, more bargains at the big-box store, more airplane trips and exciting experiences, more and longer life, and settling for less is failure.

In 1972 some very smart people who called themselves the Club of Rome projected when the growth of industrial consumer society might hit the hard resource limits of our Earth home.[13] In the fifty years since, these projections have stayed more or less on track, with the additional complication of climate chaos destroying things faster than expected.[14] Using our Earth home like a warehouse store was never going to work.

Story of interdependence: Limits are real and must be respected. More is not intrinsically better. And less is not failure.

This is a story we do well to embrace so that we can face a future that contains inevitable loss. Living with less is not failure. It is a vital part of respecting the limits of the nonhuman world.

At the individual, community, or governmental level, living well with less is better done by planning and managing a transition to a simpler way of life, rather than by using everything up and then scrambling to survive. People embracing this story say, "Collapse now, and avoid the rush." We can find measures of meaning and success that do not require material wealth.

—◦◦◦—

Story of disconnection: Independence is a virtue, especially financial independence. Don't be reliant on others if you don't have to.

Independence is an illusion. What "independence" and "not reliant on others" mean in practice is procuring enough money to insulate yourself from direct interaction with those who provide you with what you need to live. Modern industrial living means entanglement in a vast web of long supply chains including shippers, miners, manufacturers, low-wage workers with few rights, and the systematic destruction of our Earth home. Only through the application of money and the destruction of the nonhuman world at a vast scale do we keep this web invisible and producing what we use to live comfortably. In other words, declaring yourself independent of limits comes at the cost of somebody else's abuse and bondage. And now the damage begins to encroach on those who believe they are independent.

Story of interdependence: Life is always interdependent. Only by recognizing the reciprocal relationships that sustain us can we begin to live in ways that would preserve our Earth home instead of undermining it, if that is still possible.

Let me be clear that I don't know how to honor reciprocal relationships instead of just paying money for things. For me, it is an aspirational goal. But that goal certainly invites me to engage the world differently. I have realized that if I attended to all of the things I consume in daily life, I would trace a web with thousands of strands touching the lives of workers and places I will never visit. This interdependence complicates and enriches life. Relationships do that.

———

Story of disconnection: Progress is good.

Oh, how I have loved this story! It celebrates all the wonders of modern science and technology. Only now I am starting to understand the costs. And I never understood that traditional cultures have wonders of their own.

Story of interdependence: Progress is a mixed bag. Right now, progress is killing us.

Every new technology has a downside when our relationship to our whole Earth home is considered. Even technophiles are scared of what artificial intelligence may do to us. The mining, energy usage, and exploitation required to bring us the fruits of technological and economic progress are killing us and our Earth home.

———

Story of disconnection: Relating to nonhuman entities as we should do to persons, with respect and reciprocity, is antiscientific, primitive, anthropomorphic, and unacceptable.

Meanwhile, corporations have the rights of people (in U.S. courts of law) and nonhuman entities have no rights. Many traditional cultures speak of animals, plants, places, and other nonhuman entities as if they were human, with agency (the ability to choose and act) and other human-like qualities. This is how people know how to be in a respectful, reciprocal relationship: treat the other party in the relationship, human or nonhuman, as a person. People skilled at two-way relationships with the nonhuman world have traditionally been seen as wisdom carriers, whether for healing arts, rituals to create identity and meaning, procuring food successfully and sustainably, or planning for the future.

To proclaim the nonhuman world inert and valued only for study or exploitation requires suppressing these kinds of respectful relationships and the people who rely on them. So moderns have dismissed or ridiculed these relationships. Admittedly, much relating to the nonhuman world that goes on in my culture is appropriating traditional cultures or selling fantasy cures. We no longer have reliable wisdom keepers to instruct us or hold us accountable for these relationships. We have much to learn.

Story of interdependence: Cultivating respectful and reciprocal relationships with the nonhuman world is a way to live with integrity and respect for our Earth home.

Let's try it, even if we feel silly. A surprising number of people I know have ongoing mutual relationships with trees and animals and special places. They just don't talk about it openly; that's not allowed.

I invite you to experiment, and keep an open mind, about ways to be in relationships with the nonhuman world. I listen with wonder at the lore of traditional people that assumes intimate and impactful relationships with nonhuman entities. The Rights of Nature legal movement is another hopeful approach. Practicing spiritualities that honor nonhuman entities is another approach. More approaches will appear if we don't censor them.

—◈—

Story of disconnection: Food is fuel, nutrition, and entertainment. It comes from the grocery store. How it comes mostly does not concern us, but please tell us a nice story about an organic farm somewhere. Develop a taste for exotic things. It's fun.

How incredibly privileged I am, to have been able to take this attitude toward food.

Story of interdependence: Food is life. All food is or was alive, sacrificed in order for us to live. Treat those lives with reverence, and protect them so they will continue to enable us to live. Grow your food, or know who grows your food, and support those people. Support the soil, the pollinators, the watershed, and the seed stock.

Growing a garden teaches the interdependence of life. As crops fail due to climate and social disruption, and supply chains wobble in response to political disruptions and fuel prices, having a garden also means contributing to human life in a real way.

———

Story of disconnection: Death is bad and wrong, and should be postponed by any means, hidden, and where possible, avoided.

While death has always challenged humans, our success in avoiding facing it is a modern turn, tied up in the capitalist story of unending growth, and in the overinflated egos of people who are used to getting what they want most of the time at any expense. Our avoidance also comes from being out of touch with the nonhuman world, where death and the signs of death are normal. We insulate ourselves further from the reality of death by outsourcing the care of our dead and dying loved ones. This story of death being bad and wrong generates a lot of fear and leaves dying people more alone than they should be. Given our uncertain future, we would do well to change it.

Story of interdependence: Death is a normal and inevitable part of life.

By respecting and caring for those who are dying, considering and discussing our own death, and remembering and honoring the dead, we are better equipped to face an uncertain future. We can normalize grief and loss, honoring them as part of the human condition, neither shameful nor wrong. We can better accept the decline and ending of human systems. This acceptance will spare us much panic and despair in times of unraveling.

Industrial consumer society is built on *stories of disconnection*. These stories create their own inertia, whether or not we still believe them. But we can breathe life into *stories of interdependence* by telling them repeatedly, reflecting on them, and exploring ways to live them. Stories of interdependence can take root and live in our imaginations, shifting our values and our actions.

This is a sampling of the kinds of cultural stories that may frame our thoughts and our actions, along with possible alternative stories. You can probably think of more. What would you add? Which stories are failing you, and which stories might replace them?

Perspective: Pop culture stories

Daniel lives in suburban Southern California, a wellspring of pop culture in TV and movies. He works as a writer. He came to realize our predicament when U.S. politics took an authoritarian turn under Trump. He loves to watch and read imaginative fiction. He uses some of these pop culture stories to invite people to explore issues related to collapse at thisistheend.net.

Daniel longs to wake people up to the reality of collapse. He knows he can't just announce to his friends, "We're screwed, and here's why." So he started a podcast, using pop culture fiction (movies, TV shows, books, comics, and video games) as a way of messaging that meets people where they are, and invites them to think about facets of our predicament. Future fiction and fictional alternative realities have long presented visions of possible futures and perspectives on current realities.

Pop culture conventions are comfortable settings for Daniel, so he also uses them as platforms to share his perspective. He speaks on panels, using fantasy elements of pop media as launching points to talk about real issues: environmental, racial, political, and social. At the 2022 Los Angeles Comic Con(vention), he spoke on a panel entitled, "Zombies, Blips, and the Apocalypse! Why Write Stories about Disruptions?" One year at WonderCon in Anaheim, California, he discussed the time travel movie *TENET*. This movie imagines the result of climate collapse, including class struggle and generational struggle. Through it, he gently addressed the existential angst and rage that many young people feel today when facing the future. Another year he discussed the film and TV series *Snowpiercer*. He pointed out that its portrayal of how different economic classes are treated in disasters should give us pause. He challenged the audience to consider how economic inequality makes any kind of collapse a very unequally distributed experience.

Pop culture stories invite emotional resonance and imaginative reflection. They offer opportunities for conversation and engagement on difficult topics in a way that can feel less threatening than direct pronouncements or predictions about our future.

Hidden stories

To attempt to understand the world is to simultaneously re-world the world; it is to change it. There is no place to stand from which we might gain a privileged view of things. Looking is intervening.

– Bayo Akomolafe[15]

Our history is the stories we tell ourselves about ourselves. Our identity is based on those stories. That is why the teaching of history is such a loaded topic in the U.S. today. Who will control the stories that our children hear?

I never got a proper history course. I attended a mediocre Catholic school through eighth grade. In high school, I got myself into the alternative track for troubled kids despite my A grades. This allowed me to camp at Montebello Ridge in the magical Santa Cruz mountains for a few weeks each summer. In my newly minted Social

Studies class, we discussed Rachel Carson's *Silent Spring* and the Greenhouse Gas Effect ("This is going to be an issue someday…"), but we learned little history. By the time I got to college, history was optional, and I studied all science all the time. I managed to avoid getting a proper history course in my entire schooling.

I now realize that nobody I knew got a proper history course.

A proper history course gives you a foundation for understanding your identity. Not the origin myth that legitimizes the current order, but the hidden histories too. Books revealing the slaughters and forced relocations of Native peoples throughout the United States, so necessary to bust the myth of empty land for European settlers, were just being written when I was a child. Many of those stories are not yet written.

My friends who attended California public schools learned about the network of historic missions established by Spanish colonizers, with no mention of their enslavement and displacement of Native people. They never heard about the State of California issuing a bounty of five dollars per Indian scalp in 1860. The history of systematic oppression of racialized Americans in the courts and government policy since the Civil War is only recently accessible to those who trouble themselves to read it. As my church's Diversity and Inclusion book group has experienced, that history is deeply disturbing to read, even transformative. The Ku Klux Klan and sundown towns (where "colored people" were not allowed after sundown, by written or unwritten law), restrictive housing covenants, and the denial of loans, all are part of my California county's twentieth-century history. I learned this only by keeping company with activists and digging out the stories behind current struggles for racial justice. Records were hidden, erased, and denied, not in previous centuries but in 2018.

Accurate stories of ecological exploitation often include peoples' dehumanization, displacement, and even wholesale murder. These intertwined stories of systematic violence toward humans and the destruction of nonhuman entities have seldom made bestseller lists. *The Nutmeg's Curse* by Amitav Ghosh[16] tells one such tale.

Pay attention. These stories were hidden for a reason. People are fighting against the teaching of accurate history and banning books in U.S. schools for a reason. Once we teach the history of cruelty and devastation by dominant cultures, the foundational stories of industrial consumer society lose their shine. It's about time.

Nihilism as a failure of stories

When the stories of our culture give us no way to live with meaning and integrity, despair and nihilism result. Young people are most vulnerable to this failure. They are the ones who must establish an identity and make a meaningful life in these strange times. At the age of sixty, my identity has been established long since, and even great disruption has trouble uprooting it. Whoever I will be in a difficult future, I have already been enough to have a secure base, and I probably can still coast on that old identity until a new way of being comes into focus. Young people have been given models and stories for identities that, in light of unfolding collapses, seem useless. The task of making an identity and a life in the wreckage of stories that no longer feel true can seem impossible. This is when nihilism arrives. The Merriam-Webster dictionary defines nihilism as *a viewpoint that traditional values and beliefs are unfounded and that existence is senseless and useless.*

—◆—

Lucas has lived around the U.S. He is 24 years old. He is passionate about immersing himself in nature. He spent the better part of a year "outside," living in the backcountry, sleeping in his car, or camping. He can't help but worry whether outside is going to be available to him in the later years of his life. He recognizes that the world is swiftly changing and that outside might soon become something inhospitable. So he tries to just be outside as much as he can.

Reflecting on our predicament, he said, "I get scared because I am pretty young. I have most of my life ahead of me and I'm recognizing that things are falling apart now. They have started to fall apart, and they will most likely continue to fall apart. I struggle with wanting to be optimistic about not only my future but the future of my friends, colleagues, peers, and the globe. I feel like I have a healthy dash of nihilism in the way I interact with the world. That dash of nihilism, I think it helps wake me up."

Lucas has started training to become an emergency medical technician. In that way, he hopes to be useful in hard times. That choice is a source of meaning for him. I asked him to say more about nihilism. It seems he doesn't want more than a dash. "I had the idea that nothing matters and nothing that I do will enact change in any meaningful way, global change. I definitely can impact the way people are feeling: being in relationship with people definitely helps. And I see the benefits of doing things that I like and enjoying what I have access to, and living my life as I feel meets my needs, but ultimately, I don't feel like any of that matters."

Can you spot the assumptions in Lucas's reflection that led him to nihilism? "I had the idea that nothing matters and nothing that I do will enact change in any meaningful way, global change." What an unbearable burden my culture has forced upon this young man, to lead him to believe that unless he bears responsibility for the whole globe, nothing he does matters. *"You can save the world," is a failed story.* And it causes harm.

Nihilism is a valuable signal that a story you have lived by is failing you. Only don't collect more than a dash! Lucas and other young people need different stories about how to live and how to be of service than the ones they grew up with.

When there is no saving the world, what remains? We can experiment with different ways of being in the world than the ones we grew up with, ways that emphasize belonging, humility, reverence, and service; ways that respect our Earth home. No need to "get it right." At this point, I don't know what right is. But learning, caring, and serving are always great starting places to explore new identities and new ways of being.

The limits of science

It is difficult to get a man to understand something when his salary depends on his not understanding it.

— Upton Sinclair

Science tells stories too. The stories of modern science have enchanted me and haunted me. Science stories have often been presented as objective, inarguable fact, and as sufficient knowledge for wise decision-making. But the stories modern science tells are at best incomplete. They do not account for the things we value and cannot measure. They often make unsupported generalizations about complex systems. Finally, they have often been used to legitimate exploitation of the nonhuman world and to serve human greed.

My dad was a physicist. I carried on his love of science with a Ph.D. in chemistry, and I held his belief that science could reliably explain how the world worked. For many years I imagined that science and technology would find a way out of climate chaos. More recently, I have been deeply disappointed at the failure of scientific researchers to tell accurate and complete stories about the state of the planet, and the tradeoffs and limitations of their proposed technological fixes.

As a small child, I asked my dad what physicists do.

He answered, "Physicists measure things."

"What things?"

"Hard-to-measure things."

In his post-doctoral research, he measured the Fermi surface of the element beryllium, a contribution to basic science in the form of an obscure but beautiful six-fold three-dimensional figure. Later, I remember watching Dad leave on business trips to the Nevada desert.

He would come home aching and exhausted from stacking lead bricks in tunnels. He had been measuring the temperatures of underground nuclear explosions.

Stacked in the hall closet of my childhood home were mementos of those tests, each a certificate on a sheet of paper with a whimsical name and a date. Were they meant as awards to my dad for his service? So many certificates. Not until I started college did I confront him in my anti-war righteousness. I demanded to know why he had participated in the testing of nuclear bombs. In a defensive voice, he replied, "To feed my family." So it is that science sometimes leaves important measurements out of the equation, measures of human priorities and impacts beyond the laboratory.

Apart from who's paying whom, stories grounded in scientific hypotheses often fail to serve us because we lack important data. We extrapolate what we know (or think we know) to complex systems that don't fit our simple models. Our Earth home is dizzyingly complex, and our scientific knowledge is far more limited than we would like to believe. Here are a few practical examples of this reality.

Long years ago, when my husband Scott and I were setting up a home together, he warned me not to use my trusty wooden cutting board to cut raw meat. He instructed me to use the ugly plastic cutting board I detested. Its nonporous surface would not absorb meat juices. It would be less likely to harbor the dangerous germs that raw meat can spread. I had been happy not to know what I didn't know and to take my chances. Back then, I had a robust digestive tract. But I agreed to his request.

Years later a study was published that proved him wrong.[17] Wooden boards made of nine different species of wood all retained fewer dangerous bacteria after cutting meat than plastic boards. Scott had inferred that porous wooden surfaces harbor harmful bacteria more than plastic. That inference was incorrect. Choosing safe cutting boards is a simple instance of a complex and vexing question. How do we know what we know? Is what we know the best way of describing reality as it is? Or is it just our best guess, given certain assumptions? Are we able to recognize and test our assumptions? Are our assumptions even testable?

Here's another example. How many layers of cardboard mulch for how long are required to kill a suburban lawn, so it can be replanted with food plants or drought-tolerant plants and be free of seeds and roots from the former inhabitants? This question comes up frequently in my community. Many of us inherited lawns in a region that often gets three inches of annual rainfall. It seems at first a simple question that should have a simple answer. Shall we make a study and determine that answer? Under which conditions? It will depend on the temperature, the moisture, the seeds that are or aren't present, and most particularly whether the lawn is a warm-weather grass like Bermuda grass. When established, Bermuda grass can resprout from roots six feet deep.

I know of countless examples like this. A good part of what we call science is actually opinion, supported by too few observations and too many inferences. We often don't know what we don't know. And some traditional lore is science, the result of long years or even generations of observation and testing.

Perspective: One shattering detail

Tim DeChristopher, sometimes known as Bidder 70,[18] is an environmental activist, critic of the mainstream environmental movement, and co-founder of the nonprofits Peaceful Uprising and the Climate Disobedience Center, both dedicated to creating livable futures and empowering nonviolent action. An experience he had in 2008, at age 27, changed his life. These are his words from a 2011 interview in Orion Magazine:[19]

> I met Terry Root,[20] one of the lead authors of the IPCC [Intergovernmental Panel on Climate Change] report, at the Stegner Symposium at the University of Utah. She presented all the IPCC data,

and I went up to her afterwards and said, "That graph that you showed, with the possible emission scenarios in the twenty-first century? It looked like the best case was that carbon peaked around 2030 and started coming back down."

She said, "Yeah, that's right."

I said, "But didn't the report that you guys just put out say that if we didn't peak by 2015 and then start coming back down that we were pretty much all screwed, and we wouldn't even recognize the planet?"

And she said, "Yeah, that's right."

I said: "So, what am I missing? It seems like you guys are saying there's no way we can make it."

She said, "You're not missing anything. There are things we could have done in the '80s, there are some things we could have done in the '90s—but it's probably too late to avoid any of the worst-case scenarios that we're talking about." She literally put her hand on my shoulder and said, "I'm sorry my generation failed yours." That was shattering to me…

I said, "You just gave a speech to four hundred people and you didn't say anything like that. Why aren't you telling people this?"

And she said, "Oh, I don't want to scare people into paralysis. I feel like if I told people the truth, people would just give up." I talked to her a couple of years later, and she's still not telling people the truth.

Perspective: The story nobody wants to hear

Bioenergetics researcher Sven De Causmaecker[21] lives in Germany. He studied the energetics of photosynthesis for his Ph.D. project at Imperial College in London. He hoped to find new solar energy sources that could work like photosynthesis. He thought that, "If nature does it that way, why don't we do it that way, as well?" Instead, he discovered serious structural limitations of climate-related science that he did not expect.

Sven researched Photosystem II, the protein complex within green plants that harvests sunlight. Photosystem II is ultimately responsible for growing plants and plant-derived fuels (biofuels), and every biologically based energy generation scheme.

He and his collaborators quickly built up knowledge and tools to test claims about the energetics of all kinds of processes. From early on, he realized that the dramatic claims made at that time about biofuel production from algae were suspect: "We're going to be pumping oil back into the wells with all the biofuels we're producing!" He knew that the micro-organisms in use were not up to the job.

The whole system of research, he began to realize, was set up to invite such unrealistic claims. The scientific researcher must continue a hopeful narrative to get the next grant funded. "We only need ten more years of research money, and then we'll find great solutions that will save the planet." Every grant or paper that scientists write is required to contain an impact statement. "By progressing in this research, we can solve humanity's problems." These claims started to annoy him because he and his collaborators were always coming in behind with the energy balance calculations and realizing, "No, that doesn't work. The energetics don't work. Basic principles say you can't *make* it work."

Sven came to believe that while science says it is the solution, as it's set up, it's actually part of the problem. Scientists are forced to stick to a story like, "Trust the scientists, and it's all under control. Just wait for a few more years, and we have this portfolio of great solutions, and it's going to be all nice and progressive and green tech, it's going to be great." That story is not correct. The problem is not that we lack some needed technology, it's just the basic energetics of the world. We are using more energy and resources than we can generate, even on first principles. But researchers who admit this are no longer active researchers. Ours is not a scientific problem at its root. It is a social problem of resource overuse and denial.

He was surprised when he "looked under the hood" of mainstream climate prediction models cited by the International Governmental Panel on Climate Change (IPCC). They are claimed to be hard science, but they are "guesstimating" at best. The whole idea of carbon budgets is a vastly simplified concept that allows people to factor in their wishful thinking while ignoring all kinds of unknowns. Hidden in official reports are ridiculous assumptions. Sven only discovered the emperor had no clothes by digging up references in the footnotes of references, five levels deep. Dramatically important factors are just left out of predictions of global warming because we don't know how global warming works. "It's as if we are driving down a freeway and steering by looking in the rear-view mirror," he told me. That works until the first bump, or until the road narrows and takes a sharp turn, and now we are starting to see those kinds of unexpected effects on climate.

Sven tried to write papers on his findings, on the side while doing his regular research. But he found that even if he managed to get these debunking papers published in scientific journals, they did not gain traction because they don't conform to the acceptable story. They didn't get picked up by policy institutes; they didn't get cited much. He didn't get invited to give lectures or discussions. He tried to educate people and generated very little impact. Finally, he took some time apart from his regular work to put his results in an accessible form so that climate activists, not dependent on the acceptable story for their livelihoods, can better understand our predicament. If you like technical stuff, do browse his website, svenergy.info. He explains in detail why green technology isn't as much of a solution as we had hoped it would be, why carbon capture and storage don't work on large scales, and why renewable energies don't support the same types of societies that fossil energies do.

Sven wonders if anybody is paying attention. After doing this work for a couple of years with no income, he is regretfully back to work in the collapse-oblivious world.

Stories to PLAN by

If you intend to influence academic discourse or governmental policy, your efforts are helped when your stories are published in academic journals. A broadly interdisciplinary group of academics recently published some useful stories entitled "Modernity is incompatible with planetary limits: Developing a PLAN for the future."[22] Their ten one-sentence stories are meant for academic and policy discourse. Some are a little abstract for personal application, but they are still great conversation starters. Here they are.

- Humans are a *part* of nature, not *apart* from nature.
- Nonrenewable materials cannot be harvested indefinitely on a finite planet.
- The ability of Earth's ecosystems to assimilate pollution without consequences is finite.
- Energy throughput is essential to all human activities, including the economy.
- Technology is a tool for deploying, not creating, energy.
- Fossil fuel combustion is the primary cause of ongoing global climate change.
- Exponential growth, whether of physical or economic form, [will] eventually cease.
- Today's choices can simultaneously create problems for and deprive resources from future generations.
- Human behavior is consciously and unconsciously shaped by mental models [from] culture that, while mutable, impose barriers to change.
- Apparent success for a few generations during a massive draw-down of finite resources says little about chances for long-term success.

The authors are professors of physics, environmental studies, anthropology, design, and sustainability. Their "PLAN" is the launch of the Planetary Limits Academic Network[23] (PLAN). It is designed as a community of scholars (both inside and outside the academy) who understand the complex nature of planetary limits and are open to collaboration on research addressing the predicament our civilization faces. As with many such efforts, its impact to date is modest.

This list is a great tool to pull out when people start talking about fixes that are based on failed stories. Don't expect to convince anybody, though. Remember, *it is easier to deny what one is seeing than to deny the stories that define one's reality.*

Choose your stories wisely

New stories as antidotes to stories of disconnection may have their own issues. Do not expect any one story to be a perfect template for living.

Here is an example from my own experience. My mother longed to become a doctor. She was told she could not. As a woman, she could only become a nurse. So she did, but she never forgot her dream. She told it to me, and her story invited me not to settle for women's roles. I was able to choose two professions that were historical men's work: first scientific research and then Christian pastoral ministry. At the same time, I internalized an unspoken part of this liberating story: that traditional women's work, tending home and family, and any work that is done without a paycheck, is less valuable than paid work. It was certainly not the work on which I wanted to build an identity. My husband Scott and I had a standing joke for years that we needed a housewife, meaning somebody to do the work that was essential for our feeding, comfort, childrearing, home management, and managing our schedules. This was exactly the work that we couldn't bring ourselves to value because it wasn't paid and didn't constitute a career. I have been part of a culture that, in freeing women for fuller engagement in industrial consumer society, devalues the essential work of tending and caring, and the people who do it.

Stories that are offered as remedies to unfolding collapses can also bypass the real challenges of our predicament. People offering these stories usually mean well, but the stories just don't hold up to our complex reality, or miss important parts of it. A short form of some of these problematic stories is given below.

—◦◦◦—

People are looking for solutions to "save us." Some of them are worth doing. Most of them are problematic. I don't believe any of them will save us.

Story: Technology will save us.

Please help cushion our fall with technology. But don't expect it to save us. As Sven uncovered in his research, every technological fix has limitations and environmental costs. Our predicament comes from biophysical reality.[24] The people promoting technological fixes don't want to tell you their limitations; surprisingly often they don't even bother to notice them.

Story: Moving out to the country and living off the land will save us.

If you are called to do this, do it, and I wish you success and fun. Be mindful of how much of the nonhuman world you must exploit to create and maintain your remote home. Fossil fuel and lumber are needed to build your homestead and transport yourself and your supplies back and forth over the prairie. And have a fallback plan for local disasters. For many reasons, this approach can't work for very many people.

Story: The second coming of Christ will save us.

It may save you, but it won't save me. I am not a very orthodox Christian. I am not interested in stories that only help the chosen. A firm belief that you will be exempt from future tragedies is a great comfort, if you can manage it. It may also prevent you from caring about the fate of those you believe are not exempt.

Story: The evolution of human consciousness will save us. Or, The Great Turning will save us.

Mature, wise, and generous humans have always been gifts to human communities. I hope as many of them as possible will be on hand to guide and love us through troubled times. But to imagine that most humans will achieve enlightenment? Signs do not point that way. Scared people are seldom at their spiritual best. Enlightenment is not contagious. Nor is it hereditary. Even if all of us were to achieve enlightenment, much loss is already baked into our planetary system.

—∾∾—

Some stories invoke simplistic assumptions about humans or the nonhuman world.

Story: People are innately good, and bad things happen because of trauma / capitalism / technology / patriarchy / (or your scapegoat here.)

People have always been a mixed bag. Cultures we admire usually have, on closer inspection, some features that alarm or repulse us. Traditional wisdom teachings we admire were required to direct people in life-giving ways. Traditional people were and are just as capable of causing harm as modern people, though on a smaller scale.

Story: Humans are innately destructive. The world will be better off if (or when) humans become extinct.

People have lived in sustainable relationships with the nonhuman world at various times and places. In other times and places, we have not. Perhaps we will again.

First Nations author and literature professor Stan Rushworth responded to this story as follows: "I hear a lot of people saying, 'Well, the Earth will be better off without us.' That's kind of like saying 'Mom's going to be better off without her kids.'"[25]

Story: Nature is fundamentally good, or was until we came along and screwed it all up.

The nonhuman world is amazing and complex. We may judge it as good, but natural disasters, diseases, predators, and poisons have always afflicted humanity. We are not the first creatures to create a planetary extinction event.[26]

Story: Things were better before modernity. We need to recover a lost utopian age and way of living.

For any given age and culture, some things were probably better for some people, and some were not. Many other cultures have done far better at living in harmony with the nonhuman world than ours, but that does not make those cultures utopian. We cannot bring back the past, but we can learn from it for the future.

Story: We can separate ourselves from the forces that are destroying the planet. We can be pure.

Purity and separation are illusions. Seeking purity is unnecessary if we stop demanding perfect morality, health, or security, and stop assigning guilt and blame. As for separation, the planet is smaller than we realized. Few of us know how to live without modernity. Where would we go on a small and wounded planet? Who would we displace?

———

Other stories assign us impossible work. They give us agency and purpose until we burn out.

Story: Our job is to save the world or at least to save something important.

This particular story underlies a lot of activism. At first, it can function as an exhilarating rallying cry. We can do this! We *must* do this! But as the struggle goes on and on, and the losses add up, this story begins to feel heavy and disheartening. We will either delude ourselves or realize that we are failing to save the world, every time. I treasure activism, and I invite people to undertake it with more realistic aims: to help a human or nonhuman community, or to proclaim the value and dignity of that which may be lost, or because it is for you the right thing to do.

Story: We must tell people our problems are fixable, or they will give up trying and despair.

This is the stance of many scientists and writers on climate change. To me, it seems like lying to a patient with a terminal illness. You may not feel qualified to deliver the bad news or to care for a person hearing the bad news. Still, sending people in frantic search of fantasy cures wastes their time and effort when they could be savoring and making the best of the life they have left. It also destroys your own credibility, as leaders in my culture seem to have already done with many of our young people.

Does this story actually mean, "I must tell people our problems are fixable, so I can deny my own fears"? Then your challenge is clear. Begin the work of facing your fears. Or does it mean, "I must tell people our problems are fixable so I can continue to receive a salary or recognition for pretending to do the fixing"? Then your dilemma is clear. You must find a way to work with integrity.

—◊◊◊—

When we mean well and see no way to live with meaning and integrity, be assured we are caught in a failing story. There is always a way to live if our guiding stories reflect reality and our values. We may face grief and personal upheaval, but we can still live with meaning and integrity as the storm descends.

Summary and reflection

- Facts and lists seldom change our minds or call us to act. Stories do.

- A story can invite a way of being in the world. Some of our stories, underlying whole economic, social, and political systems, worsen our predicament. These *stories of disconnection* are failing us.

- *It is easier to deny what one is seeing than to deny the stories that define one's reality.*

- If we want to live differently than the dominant culture, we will need to claim different stories.

- I have offered an assortment of *stories of disconnection* that I believe are failing us, and *stories of interdependence* that might take their place. How do these stories land for you? What stories would you add?

- As Daniel shows us, pop culture stories can be imaginative invitations to expand peoples' understanding of our predicament.

- People are avoiding accurate stories (historical accounts) of colonialism and racism, and banning books in U.S. schools, for a reason. Once we know these stories of cruelty and devastation, we tarnish the shining foundational stories of industrial consumer society. We can no longer claim unalloyed virtue for that way of life.

- Nihilism is a sign that a story you have taken to heart is failing you. *"You can save the world," is a failed story.* Young people are being asked to make an identity and a life in the wreckage of stories that no longer feel true. So they are especially vulnerable to nihilism.

- The stories that science tells may seem like objective facts. But much of modern science is actually inference or is warped by the constraints under which scientists can receive funding and publication.

- Scientific findings that don't conform to cultural stories of progress and "fixing" the environment seldom get funded, published, or discussed.

- Sven discovered that on first principles, most of the climate and energy fixes being researched can't work at scale. But nobody wants to hear this, not even scientists.

- As Tim DeChristopher learned, scientists can relay facts without telling the truth. The IPCC (International Panel on Climate Change) has been doing this for at least fifteen years now.

- Alternative stories, meant as antidotes to stories of disconnection, may have their own issues. Please choose your stories wisely.

- When we mean well and see no way to make a life of meaning and integrity, be assured we are caught in a failing story. There is always a way to live well. Find a story that offers you a way to live well.

CHAPTER THREE

Stories for courage in dark times

In this chapter, you will find stories that use powerful metaphors to describe our predicament. They don't have happy endings. Happy endings do not ring true to me. Nor do these stories have ultimatums: "These dire things will happen unless you..." Instead, these stories acknowledge the darkness of what we face and offer constructive ways to live with the unknown. I believe that we need courageous stories like these to face deep losses without falling into despair.

The first story: Our addiction. Why can't we just stop using fossil fuels? Because we are addicted to them. Addiction is incredibly tough to let go of.

Perspective: Doom and post-doom. A purposeful perspective on endings from Michael Dowd and friends.

The second story: Hospice situations. What do we do if we can't fix our predicament of fossil fuel addiction and overconsumption? Give hospice care.

The gifts of death. A list, inviting us into a cosmic perspective on the inevitable.

The third story: Our crumbling cliff. My favorite story offers the metaphor of a crumbling sea cliff for the losses we face. Those on the cliff don't want to look down. To their surprise, there is life at the tideline after collapse, though it is deeply challenging.

Perspective: Dominica after the hurricane. Jessica shows that loving responses in troubled times are possible, from a community that expects both trouble and mutual aid.

The best stories. What makes a story work for shaping our lives?

The first story: Our addiction

We admitted we were powerless over fossil fuels and overconsumption, and our lives had become unmanageable.

– Step One
(from the Twelve Steps of Alcoholics Anonymous,
modified for our drug of choice)

We in industrial consumer society are addicted to fossil fuels. We are furious at our pushers, the oil companies and their investors, for destroying us. But how would we live without the liquid fire they provide that powers almost everything we do? Some of us have realized what should be obvious by now: it's killing us quickly, not slowly. Sadly, that realization does not better equip us to live without it. If the whole industrial consumer system quit "cold turkey," suddenly and completely stopped using fossil fuel, many of us would starve and freeze in the dark. So we make bargains with ourselves and promises about cutting back. This is a familiar strategy among addicts who want to be functional and still use their drug of choice. We'll wait for the technological fix. We'll taper down on our use of the addictive substance. We'll stop in this many years…

The essence of addiction is this: you need more and more to get the same effect. We call that effect "economic growth." So mostly we don't taper down our use when we say we will. We play at switching one kind of fuel for another. We may use wood pellets, claiming those are renewable since the forests we are burning might grow back someday. We thought we should use natural gas with its slogan "the cleanest fossil fuel," until pipeline activists convinced us that was greenwashing. So we must rapidly switch to electricity, that's clean, right? Wrong. Electricity is not an energy source, merely a means of delivery. It may be clean on a sunny day at noon or in a geothermal hot spot. When the sun doesn't shine in California, our electric companies burn… natural gas! (In part.) California passes laws requiring electric cars. We don't count the fossil fuel needed to build them and to strip mine lithium for batteries, or the staggering electrical infrastructure required to charge them. We don't talk about why we need so much fuel: to live an unsustainable lifestyle based on hurtling ourselves about the landscape in two-ton piles of metal.

We have a diagnosis: climate collapse. That will take us down like cirrhosis of the liver will take down an alcoholic. Like that addict, we seek drastic solutions. Is there no liver transplant available? Or, say, a tech fix for global warming? That might buy us time, right? We don't know *how* to stop using up our Earth. Even slowing down a bit disrupts our lives to the extent that it triggers political backlash.

Sometimes an intervention by family and friends successfully guides an addict into recovery. Who might these people be for us? People with traditional and indigenous lifeways know it is possible to live with respect for Earth and her limits. They write and speak about living differently. But the addict seldom listens. More often, the addict has to hit bottom, to suffer intensely enough that they are willing to rearrange their lives to do without their drug of choice. How much suffering is enough? We have yet to find out.

Addicts do things that make no sense. If you try to remove their substance from them, they will find a way to get it, stopping at nothing. What is wrong with them? Can't they see that they are killing themselves? The addict may be in denial, as so much of our culture is. "What problem? I don't see anything wrong with this wonderful empowering lifestyle." They may admit they have an issue, but "I'll just cut down a little…" Or they may desperately want a way out, but they cannot function without their substance of choice.

When does an addict hit bottom and start recovering? When they stop digging themselves in deeper. Some addicts never stop digging until they die. My sister was among them.

Many traditional and Indigenous people, mainly in the global South, resist this addiction, though they feel its allure and its effects. So we know that not participating in the addiction is possible. Their life in relative balance with our Earth home is missing many comforts and privileges of modernity. But it often includes deep wisdom and spiritual connection with the community and the nonhuman world.

These connections satisfy needs for meaning and belonging in ways modernity cannot.

This metaphor of addiction applies not just to individuals but to our whole industrial consumer society, which is built on fossil fuels and overconsumption. Can whole societies even have transformations like that required to end addiction? We don't know what hitting bottom will look like for a given community or country, or society, though some guidelines have been offered.[27] Industrial consumer society probably cannot stop overconsumption until it dies, whatever that means. But we can take some steps that may aid positive transformation in small ways, before or after the death of our addicted system.

First, *we can name the destruction, as it unfolds, for what it is: the result of an addiction to fossil fuels and overconsumption.*

Cirrhosis of the liver is not a random disease. It is usually the result of late-stage alcoholism. Wars, economic upheavals, supply shortages, and mass migrations are not random events. They are usually caused at least in part by the destruction or depletion of the resources people need to live, or the fear of that loss. Correct naming matters. Otherwise, blame gets assigned to the innocent, and solutions are proposed that make matters worse.

Second, *we can befriend and support those who are suffering because of this addiction.*

This might mean befriending and supporting the nonhuman world, refugees, burnt-out activists, young people in despair, people in our own communities lacking the necessities of life, people suffering under authoritarianism and oppression, and more. So many are suffering because of our addiction. I can only choose a few and seek to make peace with myself when I am not able to do more.

Third, *we can experiment with steps toward living simply and respecting others, especially the nonhuman world.*

Not because that will cure our addiction, but so we can begin to envision a way of living free from addiction and full of reverence and connection. This step is the most fun by far.

Perspective: Doom and post-doom

If you admit that the future holds great loss, you may be labeled a "doomer." But facing reality with compassion and courage is not about doom. These definitions from Michael Dowd and friends at postdoom.com make that difference clear.

Doom
definition

- A normal feeling of disgust or dread upon realizing that technological progress and economic growth are the root of our predicament, not our way out.

- A name for the anxiety and fear called forth when living in a corrupt, dysfunctional civilization causing a mass extinction.

- The mid-point between denial and regeneration... with or without us.

Post-doom
definition

- What opens up when we remember who we are and how we got here, accept the inevitable, honor our grief, and prioritize what is pro-future and soul-nourishing.

- A fierce and fearless reverence for life and expansive gratitude— even in the midst of abrupt climate mayhem and the runaway collapse of societal harmony, the health of the biosphere, and business as usual.

- Living meaningfully, compassionately, and courageously, no matter what.

The second story: Hospice situations

We're all just walking each other home.

– Ram Dass

Hospice care can be a metaphor for how to face our predicament. The predicament we face includes the dying of human and nonhuman systems. What, exactly, will die, and when? We don't know, and we don't need to know the details. We do know that species, ecosystems, people, and envisioned futures are dying. For some, democracy, economic security, or industrial consumer society with all its perks are dying too.

Most people have not yet accepted this "diagnosis." Can we accept that a certain amount of collapsing of the human and nonhuman world is inevitable? Admitting that something is dying or ending allows us to stop prescribing futile fixes. Instead, we can relax into that process of dying or ending. This paradoxically means taking action to live as well as we can manage in the time we have left. This is the art of hospice care. The story of hospice care of the dying allows us to see that there are better and worse ways to collapse.

Accepting this story means coming to terms with the limits of our personal power. It means coming to terms with the reality that losses are underway that we cannot prevent. This flies in the face of familiar stories about growth, progress, and human power. Yet to those who work in end-of-life care, it sounds familiar.

The metaphor of hospicing the planet has been explored by several people. Dar Jamail has reflected on climate collapse in light of witnessing and caring for his dying friend.[28] The book *Hospicing Modernity*[29] by Vanessa Machado de Oliveira explores this metaphor through a decolonizing lens.

I am more comfortable with the end of life than most people in the U.S. As a pastor, I have accompanied families during illnesses, dying, and grieving. I also spent a little time working in a live-in hospice. In that place, accompanying people who are actively dying is both reverent and routine. My main function in these roles has been as the nonanxious presence. I have assured people who were dying, and those who cared about them, that the process of dying is a normal part of life, and that living can and should go on in the middle of dying. This understanding contradicts our stories of perpetual growth, more is better, and independence as a virtue. Few can care for themselves at the end of life. Dying means slowing down, often valuing the relational and the spiritual over the physical, and being utterly unproductive and sometimes unlovely (physically).

It is a deep privilege to accompany dying people and their loved ones because dying people often understand what matters and what doesn't. Their profound and difficult work inspires reverence. Dying people sometimes are willing to examine their regrets and their resentments. When they do, they often find that their old perspectives fall away, aiding a peaceful leave-taking.

Some people who are awake and aware in the days and hours before death see loved ones who have preceded them in death at their bedside. This is common knowledge to hospice carers. Such visions and conversations are considered normal, not delusions, by professionals in end-of-life care. I call them "the Welcoming Committee" and I like to guess who might be on that Committee for each of us. Don't worry if you can't think of anyone; some people who have died always made it their business to welcome total strangers. Obviously, I believe that death is not the end of us.[30] This belief gives me a great deal of comfort, and my belief sometimes gives others comfort too. Choose your afterlife. You will always be part of the metabolism that is our Earth home, and you will be more than that if you trust those bedside visitors.

You might be thinking of the possibilities of war, mayhem, brutality, the zombie apocalypse, or other horrors that could accompany or accelerate our collapses. Yes. Things like that could happen, and some probably will. Yet the more we as a culture fear dying and collapse, the more we will unintentionally *create* the cruelty that people perpetrate when they are afraid. People commit all kinds of aggression when they attempt to defend against what they fear. So please, become an amateur collapse hospice carer. Help show people that they can face our decline by loving and savoring life more, not less, as options get fewer.

Most people in my culture are not used to dealing with death, insulated as we are from the care of people during the dying process and after death. As a result, most of us are unfamiliar with its typical progressions. That unfamiliarity can cause fear. It can also cause misunderstandings. I don't believe we can know the progression of the collapses of ecosystems and human systems either. It is wise to try to come to terms with that uncertainty rather than to make matters worse by making inaccurate predictions.

In 2015, my 79-year-old father was receiving hospice care for advanced Lewy Body Dementia, the same disease that afflicted comedian Robin Williams. Dad had been struggling in a board-and-care home in California while I lived across the country in Virginia. One day my sister reported that he was "much better now." He was breathing deeply. She even sent a video to my phone. I am so glad she did. I recognized from my hospice work those long rattling breaths: Cheyne-Stokes breathing, common in the last few days of life. I hopped on a plane the next day. With the support of skilled round-the-clock hospice carers, I was rested and could hold vigil on the night of his death two days later.

Once medical providers determine that a person has become eligible for hospice care, the patient and their family usually want to know how much time they have before death. It varies greatly. Many people with late-stage cancer are put on hospice care only when they "crash." Suddenly, signs of active dying force the realization that they have only days before their death. Hospice teams find this frustrating because they can offer so little at that point. Yet the family often finds that modest offering very helpful, because most people in my culture are completely ignorant of the dying process.

Predicting the time course of chronic heart or lung illnesses, or the frailty of advanced old age, is more difficult. People can sometimes ease into a stable state of health even with severe disabilities. Sometimes, these people even "graduate" from hospice, at least temporarily. They may hang on to life for a couple more years.

So how long does your government have? Your savings? Your food supply? These are complex systems, and you are wise to take any predictions with a big grain of salt. We do not have enough experience to say for sure.

Further, what looks like the end to one person may feel normal to someone who has faced previous collapses. People can cling to life in circumstances they would have once thought intolerable. Like a person, a system with severe dysfunctions may sometimes hang onto life for a long time. Being with dying people has taught me to accept this decline and disability—after some inner work. I don't have to like or enjoy it, but just accept it as a part of life that, with love and care, can still be worth living.

When a person is put on hospice care, or even when they are declared to be near death, some family members may not accept the situation. They are upset that their loved one is "admitting defeat." "We must not give up! Surely something can be done!" Does this sound familiar to you in the context of our predicament? This attitude encourages futile treatments that torture a dying person. It can forestall any meaningful goodbyes and prevent humane care. On average, accepting hospice care prolongs the lifespan compared to refusing it.[31] More to the point, it improves the quality of life.

Alternatively, a family member may want the uncomfortable process of dying to be over as soon as possible. Watching a decline and wondering, "When will it end?" is not fun, but uncertainty is part of the journey. Life does not accommodate our plans. The analogy here is to those who expect near-term human extinction. Allowing a variety of possible scenarios is prudent, but expecting that "the end is nigh" is not helpful. The challenge is to hold the tension of not knowing. This uncertainty means accepting the possibility of near-term death or disaster. At the same time, it means serving the human and inhuman world as if things will hold together much longer than we expect.

Let an impending ending, whether it is of a human life, an ecosystem, a species, or a human system, be a chance to cherish what is good in what is ending. Trust that in time you can make peace with what you regret or resent about that ending. Let endings be invitations to clarify and live your values. As with the dying of individuals, let it be our goal to show dignity toward, and compassion for, whatever is ending.

what I learned
from having been told to get my affairs in order

to do my best to live the BOTH-AND

the tidal waves of despair AND hope
the trying AND failing AND getting up again
the burning love AND the excruciating heartbreak
the not knowing what the next day might bring

AND the gratitude for having woken up another morning

because all we know is that right now - we ARE here
and right now - we ARE alive - and we are NEEDED

doing what we can - the work of each moment
in all our differing expressions of deep adaptation

– Daniela Muhanshim Herzog[32]

The gifts of death

The gifts of death are the atoms within our bodies, created in the hearts of dying stars.

The gifts of death are our food; all food must die to give us life.

The gifts of death are space for new creations: space for a new generation, space for a new species, space for a new world.

The gifts of death are compassion and tenderness in the face of loss and limits.

The gifts of death are learning and acceptance of the limits of our own power.

The gifts of death are awareness: savoring the mystery and beauty and preciousness of life.

The gifts of death are joy and sorrow, laughter and tears.

The gifts of death are lives that are fully and exuberantly lived, then graciously and gratefully given up.

Adapted from a litany by Connie Barlow[33]

The third story: Our crumbing cliff

What looks like apocalypse in prospect often feels more like grim normality when it arrives in the present.

– David Wallace-Wells[34]

The following story best conveys my understanding of the way our communities are collapsing. Comfortable industrial consumer society is being lost in pieces, large and small, like a crumbling cliff.

———

I live in a condo at the edge of a gorgeous seaside cliff. The sunsets are breathtaking. The sea breeze is delightfully cooling on a summer day and not cold even in winter. Far below, the sea crashes against the cliff face.

I grew up nearby fifty years ago when a broad sandy beach lay below us. All that sand is gone now at high tide. Jagged rocks still jut out of the surf, but the shoreline is now sheer cliffs. Sometimes you can see the crumbled skeleton of a building that has slipped into the sea. When storms strike, waves crest the cliff top.

Nobody maps the shoreline. Nobody wants to know. Nobody is keeping count of how many people and homes are falling into the sea. In truth, there is no such thing as inland anymore. We are all living cliffside now, and more and more of that cliff is crumbling. Why does it crumble? We don't examine the causes. We look away. We whisper among ourselves, but nobody wants to acknowledge that they, or their children, might be next. We shore things up, and perhaps we move inland—well, we can try. What we thought was stable ground has recently been discovered to be riddled with cracks and cliffs, and crumbling at an alarming rate. The wealthy have bought up all the stable land. That land is probably only stable for now.

To us cliff dwellers, life at the tideline is unimaginable, intolerable, and unmentionable. So we have automatically retouched signs of life at the tideline out of our landscape paintings and our sunset photos. Our children are sick with fear; their clear eyes see rockfalls from landslides, and their sharp ears hear the surf pounding below, the very things we have schooled ourselves never to admit. We tell them all is well, and they must work hard to buy their inland security, a security that no longer exists. Then we wonder why so many of them are desolate, bitter, or even suicidal.

We don't allow ourselves to admit that life persists at the tideline. People often survive the fall. But their home is gone, and the climb back up the cliff is too steep for most.

Tidelines are constantly in motion. What appears to be dry land is flooded at high tide. What appears to be ocean is high and dry later in the same day. Tidelines take the brunt of storms, pummeled by wind and waves. The storm ends, leaving debris piled high or landmarks washed out to sea. So how do people live at the tideline? Brave souls lash together debris from whatever they can find and make rickety rafts and damp-floored shanties to call home. Storms are life-and-death affairs, and even high tides can be devastating. At the cliff top, patrols are on the lookout for tideline dwellers who climb up and try to make homes in our hedgerows and parking lots. They are imprisoned or booted back down the cliff.

I first saw the tideline when a few friends fell off the cliff but kept in touch. (What they did was lose their homes.) They have cell phones down there. What they don't have is shelter from storms, relief from the heat, or medical care. I stopped auto-retouching my photos and discovered brave souls floating in their makeshift rafts, like diamonds across the surf. Perhaps I will become one of them someday. Maybe my adult son will. I pray for their creativity in learning to live at the tideline because we will need to learn from them. Once, I imagined that if the cliffs I call home gave out, I could move to another locale. I was not paying attention when the captains of industry were undermining cliffs all over the world to bring me bargains, and themselves fortunes. It's all crumbling. No place anywhere on our Earth home is safe from the effects of climate disruption.

The undermining of cliffs continues at a scale that I can barely imagine. I want to blame the bought-out politicians, the greedy CEOs. But they have our help. We on the cliff top don't know how to live without undermining the foundations of our own homes.

Now you know why the photos are all retouched. Not seeing the people who live at the tideline is required. Not documenting the destruction of cliffs, and the pushing of people off cliffs, whether purposeful or mindless, is required. Otherwise, the happy story of our happy cliffs would be revealed as the lie it is. Our pride in our way of life would be a joke. No wonder the children are miserable.

I am old. I may not last long when it is my turn to live at the tideline. But I want to learn how they do it, those who have been hanging on and making a life there for years, or for generations. I want to honor them. But more: I want to learn their ways of surviving ongoing challenges and losses, of metabolizing grief, finding joy, and creating homes, sustenance, and community out of whatever serves. And I want to find healing. Not for our cliffs; it's probably too late for that. But for our souls. How *would* we need to live to stop undermining our cliff homes? Or how might we rebuild them, stone by stone, the work of generations?

I don't know what is in store. But I trust that life at the tideline is worth living. Life has often been precarious for humans and nonhumans alike, and still worth living.

Who knows how to live at the tideline? What stories do you tell there? What is different for you who have lived there for generations versus the newly displaced? How do you cope with uncertainty, loss, and the threat of loss as a daily onslaught? A traditional Black American phrase captures the perseverance, resourcefulness, and resilience required, "making a way out of no way."[35] People like me have much to learn from people who have made a way out of no way: Indigenous people, Black Americans,[36] refugees, and survivors of all kinds of devastation.

Finally, I want to find healing of my community's values so that when it is our turn to fall into the sea, we come up in the surf reaching out hands to one another, creative, courageous, and compassionate. We can build rafts together, living and dying together, metabolizing grief and finding joy, and creating a home for the time we are given to live.

Perspective: Dominica after the hurricane

Jessica Canham was born and raised in Canada. She has lived on the small mountainous Caribbean island nation of Dominica for many years. She and her partner have built a retreat center there. Jessica spoke about her experience after Hurricane Maria hit Dominica in 2017, to me and also in an interview with David Baum on his *Collapse Club* video series.[37] She was one of the few noncitizens to stay through the suffering and rebuilding after the storm. Here are her reflections:

> After Hurricane Maria devastated Dominica, we were on the front page of every magazine and corporate media all over the world for the first time, little Dominica. And most of the headlines were: "There's nothing left. It's gone." Even family were calling me and saying, "There's nothing left. It's gone. What are you still doing there?" But of course, there was a lot left. There were all of us, the people, and our will to survive, to recover, and to take care of one another. There was the natural world. When you look at devastation and collapse from the outside, there's a natural tendency to feel hopeless, to feel that it's all gone. But that's discounting our human instinct to survive. I believe our most important asset as humans is our ability to build these strong communities and these strong bonds and networks. We're incredibly creative and adaptive creatures.
>
> At first, it was just a shock to the system. Your whole world, the world as it was, was over. I was just trying to get my head around what was left here. It was collapse for us. It was the collapse of everything we knew. How are we going to feed ourselves? How

are we going to even get to town? We couldn't even get out of our driveway because it was just debris everywhere.

The event itself was shocking, but I think the hardest part was the recovery after. It was emotionally and psychologically really, really challenging. And it was also physically challenging. With the lack of food, everybody kind of lost 20 pounds during the first weeks until the rations and supplies started coming in from other islands. We were literally eating what was on the ground. Everything was closed. So it was really, "How do we survive?" Well, we had to turn to each other. We had to turn to our community. And we had to share what we had. Fortunately, we had crops like pumpkin and a lot of potatoes. We could survive on those things for a while. But everything else was rotting. And, of course, there was no refrigeration. There was no way to store anything. People couldn't go out fishing because their boats were all destroyed. We were in survival mode.

I had the choice to leave. Most of the ex-pats left because we did have a functioning seaport. I couldn't even think about leaving this community. You can't abandon your people. It's just that feeling of, this is what we've been preparing for, not even knowing, but this is why you live in a community, and this is what you do when there's a disaster. You figure out how to come together and pick yourselves up and rebuild. It was an amazing experience, with many positives, even though there was all this destruction and sorrow and loss. There was this feeling of, "We're a family, and we have resources, and we have agency here. We're not going to just sit and wait for the government, or wait for someone to come in and rescue us. There are so many people with skills here." We got our water supply going in two days. Carpenters in the village went around and made sure all the elderly people had their roofs put back on, finding pieces of galvanized [metal] and finding scrap wood and making sure everybody was covered. We had a village nurse who would go around doing house calls and making sure everybody was okay.

We met the farmers' group, met the village councilman, and we talked about the essential things that we needed to do, to provide, to make sure people are okay. And as we do on this Caribbean island, everybody started planting immediately. Because whenever there's a disaster, you plant food. There's just this instinctive thing, and we knew, okay, in three or four months, we're going to have some fresh vegetables. But in the meantime, we needed to make sure people were okay. Then the rations started coming in. And that was the responsibility of the village council to make sure those were distributed. Little bits of rice, little bits of flour, little tins of tuna, that's what we survived on for the first three or four months. After the storm, we also had this really dry period, which was unusual. We didn't have that beautiful, balanced rainforest to keep the temperature controlled. [Some trees remained, but they were stripped bare.] Fortunately, we didn't have any fires after that storm. So trees were able to recover. The canopy in many places still hasn't recovered. But it's green.

The prime minister of Dominica went up to the United Nations about two weeks after the hurricane and said, "We are going to become the first climate-resilient country in the world." We've had a massive push to get everybody housed in stronger homes and to rebuild roads. There's no climate denial in Dominica. We know that we're in for some rough times ahead.

The best stories

Many other stories and metaphors may shed light on our predicament.[38] I invite you to claim stories that lead you in new directions. These stories will have their own limits, but they will lead us differently from what isn't working now. They will allow us to stop making the same old mistakes. We will make new mistakes and learn in the process. We may discover a new identity that does not rely on the things that are now destroying us.

When cultural stories fail to reflect present realities, people become disoriented, bereft, fearful, and vulnerable. So, if a person has not yet found a story of belonging to replace the stories that no longer work for them, they can try on yours (or mine) for a time. The fit may be poor. If so, they can eventually discard that story for another. Still, testing it will help them learn what they need from a story in order to make meaning of their lives. That meaning offers a secure base from which to act purposefully in a changing reality.

The best stories can...

reveal something that has been hidden or forgotten

tell truths better than facts can

inspire wonder

send you on a journey, or take you home

resonate in your bones

rock your world

support your values

free you

challenge you

empower you

take root in you.

You may discover you want to hear those stories, and even tell them, over and over.

Summary and reflection

- This chapter holds no stories with happy endings. Instead, it contains three stories with courageous endings. Please take time to ponder these stories. Let them sink in. Do any of them speak to you? Are any of them right for you?

- First, **the story of addiction.** Understanding our predicament as an addiction helps explain why we are harming ourselves by not giving up fossil fuels and unlimited consumption, and how hard it would be to do otherwise. With this perspective, we can:
  1. Name the destruction, as it unfolds, for what it is: the result of an addiction to fossil fuels and overconsumption.
  2. Befriend and support those who are suffering now because of this addiction.
  3. Experiment with steps toward recovery: living simply and respecting others, especially the nonhuman world.

- Michael Dowd offers a phrase: "post-doom"– what opens up when we remember who we are and how we got here, accept the inevitable, honor our grief, and prioritize what is pro-future and soul-nourishing.

- Second, **the story of hospice.** The future will bring much death, of complex human systems that we now take for granted, and of human and nonhuman lives. Can we honor and respect the dying process, fear it less, and live well in the meantime? Hospice care for the dying does just that. Fear and denial of death create unnecessary suffering. Death, untimely or not, is an inevitable part of life that can be faced with courage and reverence.

- Reflecting on the gifts of death invites us to start naming death as an integral part of life.

- Third, **the story of our crumbling cliff.** After the demise of industrial consumer society lies a more precarious way of living "at the tideline" that more and more people are experiencing. Some have always lived this way. But most of those who have been secure "on the cliff top" have refused to see it. Now nobody is safe. Can we honor people who are living precariously, and can those of us still on the cliff learn from them before it is our turn?

- Jessica's perspective of living on the island of Dominica after the devastation of Hurricane Maria in 2017 shows how a supportive community faced a real collapse together.

- The best stories nourish you and inspire you to act. What are the best stories for you right now?

CHAPTER FOUR

Practical emotional support

Living a failed story takes a toll on your mind. Living through chaos takes a toll on your mind too. During the COVID pandemic, many people admitted to me that they were experiencing mental challenges they had never faced before, like life-changing burnout, anxiety, or depression. Almost every one of them expressed surprise and relief when I told them this was a common response to the pandemic and they were in good company. If you don't need tools for practical mental and emotional support, someone you know does. And in some future time, you may need them too as things fall apart. Please familiarize yourself with this chapter, especially the last two sections.

Being well enough. The importance of developing skills and practices for emotional and spiritual health in hard times.

Everything is awful and I'm not okay: Coping right now. Simple practices to use in the moment to help calm emotions that prevent daily functioning.

Relief from difficult emotions. A little background on dealing with difficult emotions, and an expanded list of practical tools to restore or support emotional stability.

Many voices: Dealing with anxiety and fear. People share a variety of personal approaches.

The watcher on the balcony. A practice for gaining perspective on your situation.

Getting enough sleep. Some tested strategies for getting enough sleep to support your mood and health.

Sharing calm: entrainment and co-regulation. The eye of the storm is a refuge for others. You can help other people settle and calm. You can set the emotional tone, or seek out those whose presence is settling for you.

The nonanxious presence. Facing hard times, people get anxious. One nonanxious presence can make a huge difference. That nonanxious presence could be you.

Being well enough

It is no measure of health to be well adjusted
to a profoundly sick society.

— Jiddu Krishnamurti

To embody loving responses to our predicament, I invite you to work toward wellness where you can. Living in a sick system, whether in denial or with awareness, produces all kinds of un-wellness in individuals. It may be expressed as bodily illness or chronic pain, as various mental challenges like depression or anxiety, existential despair or bitterness, or in many other ways. The root of this un-wellness could be called spiritual sickness or soul sickness, even when it has very real physical symptoms. It could be considered Earth manifesting her dis-ease in one of her appendages, that is, one of us humans. We are inescapably part of a sick system. Still, we can usually take steps to relieve some of these symptoms. We can work to avoid being an "identified patient," the weak link who feels and expresses the pain of a whole system, allowing others to avoid dealing with the underlying dis-ease we all are experiencing.

If you are already suffering enough to impair your capacity significantly, self-care is your first task. Be gentle with yourself when you cannot do as much as you want. And be gentle with me if my suggestions seem like weak tea.

Don't give up seeking and practicing ways to recover some wellness so you have some capacity for wise reflection, considered action, and fun. If one approach doesn't work, try another. Keep trying. Physical, emotional, and spiritual health are each valuable. When one aspect of health isn't available to you, another may be. I want you as fully alive as you can be, to add your light to the sparkling web of life that is our Earth home.

Spiritual health requires framing our experience with stories that work. I assume you have already found a story that might work for you about our predicament in the previous chapters. If you are asking, "What story?" please refer back to those chapters. A good guiding story will help relieve the kinds of thinking that provoke and prolong despair or chronic anxiety. If you don't have a story yet, search for one that feels true and helpful to guide you.

Please don't add to your anxiety by believing you must do the impossible. None of us are going to turn this ship around in time to keep it from sinking. Give yourself permission to mourn, permission to let go of impossible tasks, and permission to just be, to be one of Earth's beloved treasures. Let that be enough for a start.

—◊◊◊—

A theme of this chapter will be accepting difficult emotions. That means giving them space to flow and be released rather than denying them until they sabotage us.

I will be offering a variety of tools and suggestions. I understand that you may not be willing or able to apply them because of the very suffering or disability they are supposed to help relieve. I am familiar with this dilemma. Here is one approach. Find just one that appeals to you, and find an accountability partner. Commit to doing that action and reporting on your progress to your partner on an ongoing basis. Report to them on the schedule you two have established. This might seem like a big ask of this hypothetical accountability partner. In my experience, people who are giving by nature are honored to offer support in this way. Now, let's bring on the tools.

Everything is awful and I'm not okay: Coping right now

This resource was created in several versions by Australian collapse-aware psychologist Aimee Maxwell.[39]

Okay, you're reading this because you're feeling pretty awful about something related to our present circumstances. We know that feeling stormy can be rough so here are a few things you can do if you'd like to help shift your emotional response.

Are you hydrated?
If not, have a glass of water.

Have you eaten in the past three hours?
If not, get some food—something with protein, not just simple carbs. Perhaps some nuts or hummus? Brains like a bit of energy after distress; it'll help you re-ground.

Have you showered in the past day?
If not, take a shower and feel the safety and soothing of the water on your body.

Have you stretched your legs in the past day?
If not, do so right now. If you don't have the energy for a run or trip to the gym, just walk around the block, then keep walking as long as you please. Just move. This will help your body move through your feelings and if you're outside and it's safe, your bones will feel it.

Have you spoken to someone in the past day?
Do so, whether online or in person. Make it genuine; reach out and converse with someone you know, let them know how you feel and what your fears are. Let it out where you can.

Have you heard music in the past day?
Play a song at your favorite tempo, or dance/sway around the room for the length of a song. Music helps our brains find a rhythm, and melody can move emotions. If you want to stay in

the sad, play sad tunes. If you'd like to move somewhere else play some music that invites the mood you're aiming for.

Have you cuddled a living being in the past day?
If not, do so. Don't be afraid to ask for hugs from family, family pets, friends, or friends' pets. Most of them will enjoy the cuddles too; you're not imposing on them.

If daytime: are you dressed?
If not, put on clean clothes that aren't pajamas. Something you like. Give your brain something nice that helps it feel safe.

If nighttime: are you sleepy and fatigued but resisting going to sleep? Or not sleepy but you ought to rest?
Put on pajamas. Make yourself cozy in bed with the sound of falling rain or the sea or something soothing. (Try the free app White Noise.) Close your eyes, do a few long, slow, deep breaths to settle your breathing (remember to exhale fully), and stay there for 20 minutes—no screens. Try a body scan or some imagination journeying. If you're still awake after that, you can get up again; no pressure.

Here are some common things other people have said helped them deal with really big feelings:

- holding ice
- talking to someone (a family member, a friend, a helpline)
- taking a cold shower
- listening to music
- sucking on a strong peppermint
- touching different textures
- cuddling something soft
- exercising
- doing an art or craft activity

- 5-4-3-2-1 grounding exercise:

Working backward from 5, use your senses to list things you notice around you. For example, you might start by listing:

five things you hear

four things you see

three things you can touch from where you're sitting

two things you can smell

one thing you can taste

Make an effort to notice the little things you might not always pay attention to, such as the color of the flecks in the carpet or the hum of your computer.

Do you feel ineffective?

Pause right now and get something small completed, whether it's responding to a communication, doing the dishes, or organizing your things for your next venture out of the house. This is a really big thing. You can't do it all, but sometimes one small step can be enough to help you redirect your energy from worry to action.

Do you feel paralyzed by indecision?

Give yourself ten minutes to sit back and figure out a game plan for the day. If a particular decision or problem is still a roadblock, simply set it aside for now, and pick something else that seems doable. Right now, the important part is to break through that stasis, even if it means doing something trivial.

Have you over-exerted yourself lately—physically, emotionally, socially, or intellectually?

That can take a toll that lingers for days. Give yourself a break in that area, whether it's physical rest, taking time alone, or relaxing with some entertainment.

Wait a week before acting

Sometimes our perception of life is skewed, and we can't even tell that we're not thinking clearly, and there's no obvious external cause. It happens. Keep yourself going for a full week, whatever it takes, and see if you still feel the same way then.

If you feel immediately suicidal
Call the appropriate crisis line for your region:
https://en.wikipedia.org/wiki/List_of_suicide_crisis_lines.
[Dialing 988 works throughout the U.S.]

Please tell someone how you feel (for example, a family member, friend, or counsellor). Your brain is stuck and feeling hopeless and it's not able to tell you useful things. Get an outside perspective. *Ask them to stay with you until you get help.* Being with someone, even over the phone, increases your safety or if that's not possible then please contact a medical professional and tell them it is an emergency:

- call your local hospital and ask to speak to the Mental Health Team

- go to your GP or hospital emergency department—wait there until you see a doctor

- call emergency services—the police or ambulance may be able to take you to hospital

- call your doctor, psychiatrist, psychologist, or another mental health worker.

Stay here, stay safe.
You've made it this far. You will make it through.
You are stronger than you think.

Relief from difficult emotions

If you are angry, let your anger be fire
So it can warm someone chilly.
If you are grieving, let your grief be a river
So someone thirsty can drink.
If you are numb, let your numbness give you capacity
To walk in hard places and not feel hurt.
If you are broken, let your brokenness

Be what makes space for a new thing to enter.
If you are fearful, let your fear be a warning signal
That others may look up.
If you are lost, let your being lost
Make a new place and call it home.
However you are,
Keep going.
However you are,
Keep going.

– Laura Martin[40]

Anger and fear are healthy responses to unhealthy situations. These feelings demand our attention in a way that pleasant feelings seldom do. Like the red warning lights on a car dashboard, they grab our awareness, so we can act to resolve an urgent situation.

The problem with anger and fear comes when situations are not quickly resolved, and these emotions persist. Prolonging an intense emotion that was designed to address dangers of a few minutes duration damages our bodies. It can stimulate physical illness or lead to reactive nastiness (hurt people hurt people), self-harm, bitterness, despair, or depression. Persistent fear can become chronic anxiety or panic attacks.

Depression and panic attacks can disorder your thinking so that what seems to make sense to do is exactly what will make you worse. Please get help if what you're doing isn't working. In the case of panic attacks, see Dr. Scott Siskind's insightful approach to stopping them at lorienpsych.com.[41]

Current disasters and future threats are ongoing in our time, and accelerating. The provocation never ends. We must find times and places away from the red warning lights on the car dashboard. This comes in three ways. The first way is obvious, the second and third less so.

1. Finding Solace

We can take a break from the trouble for periods, and seek solace in whatever way we have discovered works for us. Here are some ways people have shared to find solace.

- Music: singing, playing, or listening to it

- Praying

- Ritual of any kind

- Dancing

- Counted breathing

- Water. Swimming, bathing, showering, or even washing your face

- Hugging, and make it a long one. If people are not available, hug a tree or lean against it for a while.

- Meditation of many kinds

- Caring for a garden, houseplants, a pet

- Phoning a friend to say hi

- Spending time with a friend or loved one

- Reading something inspiring or soothing

- Watching children play

- Any activity that allows you to get lost in its flow: cooking, art, or craft are some.

If you experience trauma-related emotional reactions, please consult the book *My Grandmother's Hands* by Resmaa Menakem.[42] He offers a deep understanding of the challenges trauma brings, as well as specific exercises that many have found helpful to "settle" strong emotions.

Finding calm and solace is a personal exploration. Your comfort might show up in unexpected places. Instead of reading more depressing news than she needs, Gwen finds comfort in digging into long articles and learning about things that expand her understanding of the human and nonhuman world. Recently she read an article that took a deep dive into what privacy means in different contexts. Before that, she was delighted to learn about the fantastic ways some animals have of sensing the world that humans do not. This learning about how the world works brings her joy.

How do *you* find solace in hard times? Make time for those practices.

2. Allowing ourselves to grieve

Grieving invites us to pass through strong, unpleasant emotions to some unknown other side. Strangely, grieving can sometimes offer physical and emotional relief from anger or fear. The feelings of anger and fear both contain a judgment, implicit or explicit: something is not right! It is likely an accurate judgment. But we often do not have the power to make that thing right in a reasonable time frame. What happened in the past can never be changed. An adaptive response is to find the grief that lies under that anger or fear: grief at a thing lost that we could not save. This may be a strange and new idea for you. I find it very healing. More on grieving is in the next chapter.

3. Shaping and guiding our thinking

Finally, it helps to notice and minimize unhelpful thinking and behaviors around unpleasant emotions. Most people have minds that behave like toddlers, wandering every which way. Like toddlers, our minds can be lovingly taught by repeated guidance to choose healthy patterns.

You can try picturing unhelpful thoughts as cars or buses that drive by, grab us, and take us for a crazy ride. With practice, we can get out of that vehicle relatively quickly to find relief. With more practice, we can watch a thought or emotion drive by and not get whisked away by it. Here are some ways to accomplish this.

- **Avoid nursing fear, anger, or blame.** Whether they are justified or not, large doses destroy our health.

- **Practice acknowledging and then redirecting unhelpful thinking when you notice it.** This includes thinking about future scenarios. "But I need to plan! To figure out what to do!" Great. Allow yourself thirty minutes a day to plan and figure out what to do. Then put your worries on a shelf for tomorrow's planning session. In between, live.

- **Avoid unnecessary situations that provoke strong feelings.** Provocations may include compulsive news-watching or reading. What news are you helped by knowing, and what is mostly a provocation to suffering? Video news is more emotionally stimulating than print; I usually avoid it. Compulsive social media use or attempting to educate people on social media may be either a move that gives solace or a provocation. If you use it for comfort, is it working?

- **Avoid having important conversations or making decisions when your emotions are high.** The stimulation of anger or fear distorts your perception of reality and takes blood and oxygen away from the judgment centers of your brain. "Let me get back to you on that later…"

- **Practice self-acceptance.** This requires a story of your role in the unfolding tragedies that allows self-forgiveness. Self-acceptance also requires a willingness to catch and redirect self-judgments as often as you notice them, and a willingness to believe that what others think of you is not your business.

- **Find a good listener** to vent your feelings and judgments. Do not expect to be able to relieve troubling thoughts and feelings by yourself. This may take some experiments and false starts until you find or coach someone into listening well.

- **Get support from professionals**: coaches, therapists, or doctors, when strong emotions or disturbing thoughts bother you *before* they disable you. Professionals can provide targeted strategies for your situation, and sometimes medication can be a valuable aid in resetting a troubled brain. Invest in yourself if you have the means. You can ask for a sliding fee scale from private therapists, or call your local government to check for

low- or no-cost options. Therapists and coaches who understand and honor climate-related issues[43] are available.

- **Any meditation practice**, counted breathing, sensory, or
guided, will do if it provides relief from troubling thoughts. If
you believe you can't meditate, please keep trying different
kinds of meditation. Do not judge your ability to "meditate
well." Rather, find the meditation style that brings you some
peace.

Which of these approaches is familiar to you? Which would you like
to try?

Forcing ourselves to suffer chronically or punishing ourselves for
the state of the world, intentionally or unintentionally, increases
suffering for all. Seeking relief at intervals, or even for a season when
you are overwhelmed, is different from avoiding difficult situations
altogether. We know the troubles remain, and we will witness and
care as we are able and called to do so. We must respect our limitations and be able to act from, and for, what is life-affirming.

Wake up into a day that you did not summon.
Go out remembering
The first name that you were ever given,
The one told from infinity and fragments of stars.
Remind others of that name.
Find something soft,
And let it find you too.
Believe in the life that lives in mud
And stories salt tells.
Translate what you can
Of the good and the near.
Know that grace is as real
As gravity.
Receive.

— Laura Martin

Many voices: Dealing with anxiety and fear

When I am afraid, I am miserable. I play it safe. I restrict myself. I hide the talent of me in the ground. I am not deeply alive—the depths of me are not being expressed... When I am afraid, I am a house divided against itself. So more than anything else I want to be delivered from fear, for fear is alien to my own best interest or, to put it positively, I want to give myself generously, magnanimously, freely—out of love. I want to be able to take risks—to express myself, to welcome and embrace the future.

– Gordon Cosby

I asked participants of the Deep Adaptation private Facebook group to share with me how they cope with fear and anxiety. Feel free to skim or skip this long list if your current strategies are working. Here are some of the many replies, listed by topic.

Explicitly addressing fear:

- I embrace it [fear about the state of the world]. I hadn't cried in many years. Occasionally now, I do and it feels very releasing.

- "Fear is the mind-killer" is one of the truest statements I have found in this life. Humans need stories to learn and the place of old fairy tales and fables often filled the cultural role of fencing the fears of the communities.

- When I learned about a fear as a child, I pushed my limits against it. So for claustrophobia, I would take a sleeping bag into the little tunnel under my bed and roll in it in a tight ball in a tight space. For heights, I would go up on a roof. For drowning, go swimming and hold my breath underwater as long as possible. For creepy crawlers, snakes, etc., find non-venomous types and play with them. For every fear I essentially did exposure therapy before knowing what it was.

- Fear has been my default emotional state since I was little. A game-changer for me has been to learn to be more accepting of it.

Adopting a specific perspective or philosophy:

- I sometimes think about my ancestors who were shot at, took part in resistance against Nazis and Soviets, ended up in camps, and had to migrate to the other side of the world. They had no electricity, a few clothes, often no shoes... My life is easy-peasy compared with theirs.

- I have community! Connection to community has enabled me to drop intense fear, to be able to risk arrest by the police, for example. It's important to say, I am a financially independent, middle-class white settler.

- Try to think big perspectives - grasping big/deep time, generations, cycles of life and death, the height of big trees and the extent of roots, the size of the Earth, the size of the solar system, galaxy, and universe.

- I remember that there have been mass extinctions several times in the past. The earth will continue. Some life will continue and will evolve. My life is insignificant in the bigger picture.

- I follow an Epicurean and Stoic path, being with what is, living simply, accepting reality, and cultivating virtue for its own sake. I have daily sanity-check conversations with a small group of friends on [the Discord app], a "philosophical fellowship."

- I allow myself the space to believe I'll be around for at least a decade. (I always have a rolling decade before me.) A long enough plausible horizon has made it reasonable to invest time and energy into making my home more of a sanctuary for respite. I also find meaning in trying to be helpful.

- I deal with the present human-made mess with...the sense of humor I was blessed with, timefulness (Geologist Marcia Bjornerud wrote a book of that name), and acknowledging impermanence (most important of all).

- I have gotten a lot of benefit from reading end-of-life—related books to prepare for a loved one who was dying. Also books by Eckhart Tolle and Pema Chodron… Also in my life (and more recently, in embarking on my anti-racism journey) I have met and read about many people and entire cultures who have experienced a total loss of everything.

- Trying to have some patience and compassion for myself seems most useful to me. I have become more appreciative of the method my body/spirit/mind chooses on its own. I dissociate.

Taking a break from bad news, or stimulation in general:

- When my head is spinning and I suspect I am going to be waking up in the wee hours, I take L-Theanine before bed. It is a (nonessential) amino acid also found in black tea. It eases my rumination, and I'm told it is compatible with psych meds. I don't take it regularly; fortunately, my head only spins occasionally.

- Withdrawing from the outside world and being still, not dealing with other people any more than absolutely necessary. A day or two in retreat usually suffice for me to emerge and function "normally" again.

- The most important thing I do is limit my exposure to negativity and media exaggeration. I read select news articles to stay aware and have curated my Facebook feed to groups that are thoughtful, creative and/or relevant to my interests.

Meditation and spirituality:

- I have been almost overwhelmed with anxiety over the last 2 months [from a variety of events.] I cope by following a deep meditation practice several times a day of:
  1. Breathing out and down into my body,
  2. Checking the tense places and breathing into them… pausing.
  3. Breathing into my mind and seeing how overcharged, busy, and unregulated is my mind, calming it… pausing.
  4. Asking myself: who has been doing this previous 3-step process?…Who is it that is aware and experiencing these sensations, feelings, and thoughts? Then connecting to this true self that simply is…the human being part of me, rather than the

human doing. Being with this energy field for a few minutes. Enjoying this me, celebrating my life. Being grateful for being here at this momentous time in human history.

- I take THC gummies and listen to doom metal while I paint and at night before bed. I've really been into some music where the lyrics speak of oneness with nature, a becoming through unbecoming, growing into something magical/powerful by divesting of the earthly body. I just try to focus on participating in the cycle of life, growing old and decomposing like a tree...

- What has offered me the most equilibrium is simple mindfulness. I also consistently acknowledge that I have never once experienced a future. I have only ever experienced a now, the present happening of this moment. I will never experience this imaginary future. It will always be now, and the response will be now as well. I have a poem that helps me ground into this present moment which I share:

> The breath is happening breathing
> The heart is happening beating
> The moods are happening mooding
> The thoughts are happening thinking
> The planet is happening turning
> The seasons are happening shifting
> The sun is happening shining
> The sky is happening blue.

Taking action, of many different kinds:

- Action! Action, love instead of fear. Action: hug my kid. Action: go out in nature. Action: breathe, watch something funny before bed, also sometimes whiskey, music, and dancing and see what's new on [social media groups that discuss constructive action.]

- Some things I do when I remember: Watch birds. Write. Sit outside. Weed the garden. Walk. Get under and up into big trees. Conscious Breathing. Feel soft textures.

- Being outdoors...I'm nature. I take loooooong walks...my walks are not only exercise but therapy, serious therapy. I walked for almost 2 hours yesterday, beautiful day. I had issues before I went, much better on return.

- The Work That Reconnects has really helped me hold space for my grief and fear around global collapse and the ecological crisis. I've been attending the workshops and I'm now training to be a facilitator for the work.

- Action takes me out of spinning anxiety. I call or write a friend, garden, cook, bake, can or preserve foods. These things shift my focus via what I would call "good and useful distraction." They don't ignore our predicament entirely but shift my focus to things I can control and that reflect my values—providing for family and loved ones physically, emotionally and spiritually.

- My approach to being with and moving through fear: several years ago, I started practicing re-evaluation counseling, which is a peer counseling practice that embraces the natural human process of releasing emotional, physical, and mental tensions through non-repetitive talking (mostly recalling memories we don't normally talk about), crying, laughing, yelling, yawning, trembling and perspiring. These are the biologically programmed ways our bodies know how to release fear, grief, anger, etc.

- I read escapist fiction or watch silly romantic comedies; I find that it clears my mind. It's so lightweight, that I don't have to think, and that enables me to breathe lightly!

- I was facing cancer surgery early this year and used my waiting time to get as healthy as possible and plan a large container garden on my deck. I started seeds indoors and created a timeline for planting what I wanted to grow. I felt such hope when things began to sprout just a few days before my first surgery that I actually cried tears of joy. After surgery, I slowly paced myself in getting each of these little wonders hardened off taking them out to enjoy the sun and breeze during the day (and get myself out for the same!) and eventually into pots. As summer came, I began to harvest my produce and felt awe that I grew and nurtured these things from tiny seeds over many months, participating in the miracle of life. I now have salsas, pizza sauce, spaghetti sauce, canned tomatoes, pickles, relishes, and jams to enjoy and gift through the winter months.

What are your strategies for dealing with tough times? What works to ease your anxiety or fear? It might help to "test-drive" some of the offerings earlier in the chapter so that you can be ready to face the next storm.

The watcher on the balcony

When I am flooded with emotion, or haunted by the ghosts of old thoughts or beliefs that no longer serve me, I am not able to participate in "the dance of life." I find myself stepping on toes, tripping over my own feet, spinning dizzily, or just stopping in my tracks, frozen by confusion or fear.

At times like these, I remember the watcher on the balcony, above the dance floor of life. The watcher on the balcony is me, but a calm me apart from the drama of the moment. This version of me has access to my best wisdom and resources, and can see the larger perspective.

Here are some of the things the watcher on the balcony has seen:

The person who tripped me looks from the balcony to be spinning out of control. He did not mean harm to me; he is lost in his own emotion.

The person who lashed out at me thought she was protecting herself.

The gut-wrenching agony of grief that left me frozen did not last. It passed after a deep cry and some journaling; the next bout of grief need not scare me.

The more I demand of myself, the more I stumble.

The despair that tells me nothing I do matters can be eased by doing something seemingly trivial that engages my body and my senses. Things that ease despair matter and are worth doing.

I cannot control what other people think of me. I can seek to understand their heart, their feelings and needs, rather than their judgments.

I am loved beyond measure, forgiven, blessed, and supported by the Sacred. I have those gifts to sustain me and to share.

I don't live on the balcony. It's a bit cold up there, too distant from the dance of life. I keep living on the dance floor, dancing a good part of the time, sometimes getting tripped or stumbling or stopping in confusion. Then I come to myself and look up to the watcher, who helps me to rejoin the dance.

The Watcher on the Balcony Practice

Picture a time when you could not enjoy "the dance of life." Imagine yourself as a watcher on the balcony. Can you gain perspective on the situation now that you couldn't see then?

If you are unable to do this thought exercise, you may want to seek a therapist or a wise soul who can guide you into this ability to take a wider perspective on your situation.

Getting enough sleep

Whatever emotional load you carry, whatever other stressors in your life, sleeping well helps you cope, and sleeping poorly adds to your woes. Worrying about not sleeping well adds more suffering. The following are some simple approaches that have worked to improve sleep. Like the other suggestions in this chapter, these approaches require some initiative and commitment. This can be hard to find when you are discouraged *and* tired. Again I suggest you find a person who will support you and to whom you will be accountable. Or if you're not yet ready to care for yourself in that way, keep a diary. Make it real; make it matter. Your well-being matters.

Visions of catastrophe don't help you sleep. Yet Aimee Maxwell assures us that it is possible to have anxiety and still sleep reasonably well.[44] Do address your anxiety, too, if that is an underlying issue. But even before you find ways to cope with those anxious thoughts, you can take the following steps to help your animal body regulate your sleep.

- Partake of about 15 minutes of natural daylight in the morning to help your body know the day has started. A short walk is perfect. Sunlight in the morning triggers your body to make the chemicals that help you sleep at the end of the day. If you have been waking up in the late morning or afternoon, set the alarm 30 minutes earlier each day to shift your sleep earlier. Brutal, I know. You will probably get less sleep at first, and you will be more ready to fall asleep at night.

- In the exhaustion or depression that often accompanies anxiety, your body gives you the wrong message: that you should rest, conserve energy, and hide from the world. Resting and hiding feed exhaustion and depression. Your animal body must move during the day to know it's daytime. Walking is excellent exercise. For me, gentle activities like walking or yoga usually relieve the fuzzy-headedness that comes with a lack of sleep. A win-win!

- Turn off screens an hour before you go to sleep; dim the lights too. This requires thought and commitment. You could set the alarm for the time when your day begins to wind down, put away electronic devices for the day, and do... wait, what do we do without our electronic devices? Being a little bored is the point. If you are unable to step away from screens, if it is costing your health, that is an addiction. Perhaps gentle tidying up, stretches, meditation, journaling, playing music... ah, I already feel myself relaxing, stepping away from the computer. Creating routines helps relieve anxiety too.

Dr. Scott Siskind's website lorienpsych.com lists more sleep hygiene remedies that are simple to try (but not necessarily easy), including:

- What happens if you stop drinking alcohol for a few weeks? (If you can't, you may be an alcoholic.)

- What happens when you stop drinking coffee? Even in the morning. Just as a temporary experiment.

- How can you make your bedroom as cool, quiet, and dark as possible? This one is key for me. I use earplugs and an eyeshade every night.

- Can you coach yourself to feel peaceful about not being asleep when you think you should be? There's nothing like anxiety or annoyance about being awake to keep you awake.

After the above remedies have been tried, congratulations! Siskind's site offers several other low-cost strategies for improving sleep, including tested relaxation routines and therapy apps, and a menu of supplements to try before seeking drugs (all of which have issues.)

—◊◊—

Why is self-care so much work? The good news is that the self-awareness needed to change habits will serve us when we try to live intentionally out of new stories that respect our Earth home and its inhabitants. The bad news is this: when we most need to change patterns in our lives is likely when we are least able. Get help! Pick a change you want to make and be accountable to someone or in your diary. Am I repeating myself? Are you doing it yet?

Sharing calm: entrainment and co-regulation

This section and the next go beyond personal mental and emotional care, showing how the sense of calm you cultivate can become a gift to others.

Entrainment (being carried along) in human interactions means one person's bodily function synchronizes to match another person's.

Because humans are so social and skilled at unconscious mimicry using their mirror neurons, entrainment is a common phenomenon once you know to look for it. Here is a simple example: try yawning in a group.[45]

I first learned of this effect when my son was a newborn. In the book *Becoming Attached*,[46] Robert Karen noted the advantages of babies sleeping with their carers, as they still do in most of the world.[47] He claimed that the steady breathing of the parent helps to regularize and stabilize the breathing of an infant; that parent and child then breathe in rhythm with one another. As I was reading the book in bed with my newborn at my side, I glanced down at that tiny sleeping being. Despite supposedly being unable to move under his own control, he had somehow attached his little hands and feet onto the side of my torso like glue. I noticed his determined huffing; he was really working at this new thing called breathing. I noticed too that, at least at that moment, he had no interest in matching my breathing. Instead, I became aware that I had been matching the rhythm of my breathing to *his*.

This illustrates an important feature of entrainment: power or size does not determine which person entrains the other. Instead, the person who can set the example with determination, no matter how tiny, can entrain others.

I want to expand the definition of interpersonal entrainment beyond its usual meaning of the physical synchronizing of two organisms to include *co-regulation*. Co-regulation is a term in psychology that describes the continual feedback of emotional systems that happens when people are together. Entrainment is lopsided co-regulation, where one person sets the emotional rhythm, and the other person follows along, unconsciously matching the stronger rhythm.

I count on entrainment when I practice Reiki, a form of healing touch. I cannot predict what my Reiki clients will experience, except for one thing. I can assure the people I serve that they will relax deeply, often falling asleep. I am deliberate in my slow and calm presence, slow breathing, and gentle and very slow movements. I'm standing up and focusing my concentration, so I'm in no danger of dozing off. I can reliably entrain the client's nervous system to mine, and they relax into butter.

An enjoyable conversation can be a flowing dance of co-regulation and alternating entrainment. But when the other person refuses to join that dance, or is unable for some reason, we feel disconnected.

Entrainment can be used to calm emotional reactions. Here's the challenge: the emotionally stimulated person is focused on the provocation, not on the calm person next to them. The calm one needs to get the attention of the agitated one. If someone wants to settle their emotions, I might invite them to take deep or counted breaths *with me*. This helps interrupt their preoccupation with the disturbing situation and allows them to join my calm. Then I might listen to and reflect their thoughts back to them, with empathy. My full attention invites theirs on me, and their attention on me invites them to entrain my calm. I may need to work on staying calm myself if their story affects me strongly. I don't listen and reflect mechanically; I resonate with their emotional expression at passionate moments. But then I return to a calm state; my quiet attention invites them to join me in calmness.

Nonhumans can also entrain us. You might call it by another name like grounding. I'll never forget the Labrador Retriever who sat with me one long night during my first year away from home at age 17, in the lonely aftermath of a heartbreaking choice. Despite barely knowing me, this dog rested his head in my lap and stared into my eyes with unshakeable love as I cried my heart out. His care for my pain brought me so much solace.

Some people find that trees or large rocks are good at restoring them to a state of peace. Earth herself, if we lie on the grass or the forest floor, can restore our emotional regulation.

Familiar groups easily co-regulate and entrain each other. The mood of the group, and even the behaviors, can run through a familiar repertoire. Part of the anxious feeling of not belonging in a new group can be the visceral discomfort of not knowing how to become entrained with that group. Even if we desire to be entrained, it is usually not a conscious process.

When a group has developed an ugly mood, you may be able to entrain them into your more compassionate state by stepping in with a powerful presence and a different perspective. You will need to be *self-differentiated*: clear on your values and purpose for acting, and able to stand up to the emotional force of an already entrained group, while still connected emotionally. You may find that using "I" statements like, "I notice…" "I wonder…" "I feel…" "I value…" will help prevent defensive responses.

Entrainment works in larger groups too. It is easier to whip a crowd into a frenzy than to invite it to become peaceful. But it is also possible to speak of sorrowful, joyful, or loving experiences and entrain listeners into those emotions. I expect to accomplish this when I preach. Most people don't seek out public speaking opportunities as I do. Still, in smaller groups, you can practice entraining others when appropriate.

If you are sensitive, you may easily be entrained into others' emotional states, whether you want that entrainment or not. With practice, you can grow in your capacity to accompany someone in their emotional state, holding space for it and resonating with it without falling into its rhythm, at least some of the time. Nobody has the emotional stamina to entrain others all the time. Allow yourself to be entrained in your turn, to receive emotional support from wise, calm, and loving people, or those who just radiate vitality and joy. A purring cat, a toddler inviting you into a game, an inspiring speaker, a wholehearted listener who doesn't judge… We were not meant to carry our emotions alone.

This entrainment business may seem more complex or subtle than you can follow. If so, remember this. Fear is contagious, calm is contagious, and courage is contagious. Who's setting the tone? Do you want what they're offering? Whose emotions are contagious? And do you want to catch them?

The nonanxious presence

Give light, and people will find the way.

– Ella Baker

Managing anxiety is a key skill in hard times. Using the tool of entrainment, we can practice being the nonanxious presence[48] to help other people find calm despite the storm. Anxiety is already rampant, so we can start practicing now.

When denial is in full swing, slow-brewing crises are not consciously acknowledged. Instead, they manifest as anxiety simmering beneath the surface, while people do their best to carry on with business as usual. Unspoken anxiety affects relationships, mental and physical health, politics, and policy. The nonverbal signals of anxiety are often interpreted (consciously or unconsciously) as anger. People find scapegoats to blame and become less tolerant of differences. The United States has been working its way to a slow boil of anxiety for a while now, and the results are ugly. Incorrectly understood anxiety starts a vicious cycle. Blame and fault are found: more to fear! Drastic action is demanded to stop those evil people: more to fear! The violence starts—more to fear!

Anxious people are prone to behaving badly, and not because of bad intent. Anxiety redirects our brains from the most complex and energy-intensive thinking, like creativity, empathy, and careful decision-making. It activates more ancient and reliable brain functions: floods of emotion, threat identification, and instinctual fight, flight, or freeze reactions. Unfortunately, news and social media make money by manipulating and amplifying our anxieties. Anxiety itself has become the cause of much unnecessary suffering and harmful behavior.

You can be the nonanxious presence by knowing and living your values and purpose within a family or group, despite provocations to panic. Pre-processing anxiety and grief, using a toolbox of relief strategies, helps to accomplish this. We can make a big contribution to the mood and functioning of a group by staying calm and connect-ed to people who can't manage their anxiety alone. This is intentional co-regulation and entrainment.

Being the nonanxious presence requires self-awareness and com-mitment. Nobody can be it all the time. Yet at crucial points, one nonanxious presence can turn the tide of a person or a group from despair, fear, or even violence to trust and care. When others see that you are keeping calm, they know calm is possible. You seem down-right attractive to them. How do you do it, they ask? The nonanxious presence is a powerful way to be the eye of the storm.

Summary and reflection

- We live in a sick system. It can make us sick physically, emo-tionally, and spiritually.

- To counter the effects of that sick system, we will need to invest in our physical, emotional, and spiritual health.

- If everything is awful and you're not okay, try the simple steps in the second section of this chapter to get quick relief.

- More tools and practices can help you get relief from difficult emotions. Which are in your toolbox?

- Anxiety is a familiar emotion in these times. People share what they do to relieve it.

- "The watcher on the balcony" practice helps us find a wise perspective on our thoughts and emotions.

- Get enough sleep. You can try using the tested strategies listed to support emotional and physical health.

- Entrainment is emotional co-regulation where one person sets the emotional rhythm, and the other person follows along, unconsciously matching the stronger rhythm. We can use it to share calm and compassion.

- Managing anxiety is a key skill in hard times. Who's setting the emotional tone? Do you want what they're offering? Whose emotions are contagious? Do you want to catch them?

- Let yourself be entrained by people who bring you calm or joy.

- When you can be the nonanxious presence, you can help other people to calm through co-regulation and entrainment. This is a powerful way to be the eye of the storm.

CHAPTER FIVE

Befriending grief

In this chapter, I offer grieving as a skill and a practice to help you metabolize the experiences and emotions of things falling apart, including losses you may not know how to name or take in.

Grieve to relieve. My personal experience with the art of grieving.

Learning how to grieve. Grieving can be a skill. It is more than stages, and it is more important than my culture acknowledges.

What we grieve. Griefs that were spoken in a Deep Adaptation community grief circle.

Perspective: Saying yes to sadness. Lisa has decided that sadness is an appropriate response to life in our world.

Normalizing and simplifying grief. A host of situations can complicate grief. And you can address them to help simplify grief and relieve suffering.

The Handoff Meditation. Practices for releasing mental and emotional burdens that are too heavy to carry.

Grief rituals. For you to do, alone or with others. They work better with others.

Cultivating joy and gratitude. Those who have grieved know what is precious and worth savoring.

Grieve to relieve

Grief is itself a medicine.

– William Cowper

When I planned my first Grief Gratitude and Courage[49] workshop for climate and ecological grief, I asked some climate activists I knew, "Do you need to grieve the state of things?" Many of them admitted they did. They admitted it in a whisper; it seemed they were not even allowed to tell each other. Some had sought professional counseling for their climate grief and anxiety; that fact too was shared in a whisper. They were activists. They were supposed to be saving the world. But they had already lost so much. I suspect that if you don't need to grieve the state of the world, you aren't paying attention.

When I notice myself getting anxious or avoiding some issue, a piece of news, or a situation in my life, I go through a sequence that is by now predictable. I reflect on the issue in my journal or with my Nonviolent Communication empathy buddy. I say, "I'm having feelings about this." Angry, scared, and confused; I continue with a list of what's wrong with me, them, and the world. I finish when I've discovered a short list of things my heart longs for, things that aren't happening. The blaming and resenting have usually faded. When I sit with my feelings and longings, without distraction, in the presence of a listener, it doesn't take long until I start to cry. My unpleasant emotions are usually a sign of unrecognized grief.

Sometimes I find myself crying at the slightest provocation. Then I realize: right. Here's what I'm experiencing, and I haven't stopped to sit with it and grieve it yet. When I finally do, sometimes a strange thing happens. At the same time I'm bawling, I'm feeling huge relief. Bittersweet sadness is out in the open. Deep caring arises, and yes, it hurts. But suppressed grief is not shutting me down or suffocating me. Grief is now looking me in the face. I can understand what's going on. I know what to do. This happens to me over and over again.

A few times, an episode of grief has been deep, raw, and long. I have passed through a space of unmaking. What was once steady ground, an assumption about how the world worked, about who or what would be present in it, vanished. The empty hole felt terrifying. Still, at the end of wracking sobs, I found some peace. Goodbye to that old reality. Hello to a new, more austere reality, with what remains all the more precious.

I'm told that I am lucky to be able to cry so easily. I'm told that stress chemicals are flushed out of our bodies in tears. I'm told that in many traditional cultures, and even in some traditional practices of monotheistic religions, the community is invited to grieve regularly and ritually. So I create such rituals for my communities. That way, I can release what is pent up in me and allow others to do the same.

Learning how to grieve

What if grief is a skill, in the same way that love is a skill, something that must be cultivated and taught? What if grief is the natural order of things, a way of loving life anyway? Grief and the love of life are twins, natural human skills that can be learned ... In a time like ours, grieving is a subversive act.

– Stephen Jenkinson

We have much to grieve in this time of mass extinction and the ending of industrial consumer society, because much that we love and have made a life around is being lost. Grief may go unnamed and unacknowledged, only finding expression through anger or depression, or anxiety. This is what I am inviting us to avoid. Because my culture offers so little support to grievers, *we will have to learn how to grieve.*

Grief is more than an emotion. It can be a practice, the hard work that we do to honor that which we love and have lost. It can be a process, a journey through loss, to discover how to come to terms with that loss. It can be a skill we cultivate to keep reopening our hearts step by step to the flow of love, pain, awe, and healing that is required for living well in hard times. Grief encompasses a spectrum of emotions and responses to the loss of something or someone we value. By participating in grief, we recognize and honor what we have lost. We give it time and attention, and *we are remade* by that loss and prepared to face what remains.

Grief may follow the rules of my culture, to be expressed as a manageable sadness through appropriate rituals, only after someone we care about has died. But grief often does not follow these rules. Either it remains hidden, untouched by the minimal rituals customary here, or it breaks cultural bounds and intrudes into life in ways labeled "drama" or even "pathological." Grief encompasses many more things than the recent death of a loved one, and it refuses to keep a schedule.

Losses travel together in our psyches; what is not grieved in the past will arise when another loss hits. The backlog can become intimidating. We may imagine that once the floodgates are breached, the grieving will never end. But the intensity almost always lessens as grief is acknowledged and expressed.

—◊◊◊—

An unscientific poll of traditional cultures reveals that a week is a typical time assigned to drop everyday activities and properly honor a death in the family. In Jewish practice, this has typically meant a week of "sitting Shiva": staying home and being fed and cared for by friends, being expected to take no responsibilities, and not being left alone. Some Christians from the Pacific islands of Samoa brought their week-long funeral practices with them when they came to Southern California. These did not last long. Often working in low-wage jobs without vacation benefits, the mourners lost their jobs when they practiced these week-long observances.

Traditional religions often include annual rituals for honoring loss. The rites and ceremonies accompanying them are ideally a chance for the whole community to grieve the various losses they are expressing. But sometimes, such rituals cultivate fear or guilt instead of healing grief. That was my experience as a grade-schooler instructed to meditate on the Catholic Christian stations of the cross, the torture and execution of Jesus.

—∿∿—

Grief may come powerfully, unsought, and take over our body and mind. This can happen when a loved one who is a huge part of our identity is lost, or when the circumstances of the loss seem unbearable. Grief can be debilitating for a time. This is normal, if deeply disturbing to those around the griever. When grief lingers long and intense, it may be *complicated grief*, discussed later in this chapter.

The "five stages of grief" named by psychiatrist Elisabeth Kübler Ross were based on her observations of dying people, not people grieving loved ones or other losses. Her list is:

- denial
- anger
- bargaining
- depression
- acceptance

Many people facing loss find these five markers an incomplete guide to their experience. They experience these markers in no particular order, and some they do not experience at all. Kübler Ross herself acknowledged that these are *not* usually linear stages people pass through, even people who are dying.

People seeking to face losses intentionally may find perspective in the "five gates of grief" named by soul-centered psychotherapist Francis Weller in his classic book *The Wild Edge of Sorrow*.[50] They are invitations to honor grief from various losses in addition to the death of a loved one.

- The first gate: everything we love, we will lose. This includes the death of loved ones and every other kind of personal loss.

- The second gate: the places (in us) that have not known love. This gate helps us to heal childhood wounds and to cultivate self-compassion.

- The third gate: the sorrows of the world. This gate is almost universal in my groups.

- The fourth gate: what we expected and did not receive. Walking through this gate can relieve anger and bitterness.

- The fifth gate: ancestral grief, the emotional wounds we carry from losses in our cultures and from past generations. For moderns, this includes the loss of connection with our Earth home and those ancestors who lived in a mutual relationship with Earth.

It is often easier to grieve together than alone. Here are some rituals and groups that support grieving the state of our Earth home:

Francis Weller and his colleagues create retreats and rituals to face grief. To find them, try searching "Grief Tending" on the web.

Collapse-aware Grief Circles can be found in the Deep Adaptation Forum events calendar, deepadaptation.info/events. Talk to me about using this format for your community.

Grief work is integral to The Work that Reconnects (WTR),[51] a set of practices usually presented in workshops or retreats, that was created by Joanna Macy and friends. Many, but not all, WTR facilitators are collapse-aware.

The Good Grief Network, goodgriefnetwork.org, offers 10-session groups for people in grief about the climate and the state of the world. Some Good Grief facilitators are collapse-aware.

Art, music, and poetry sometimes express grief better than talk. Look for lists of poems or songs related to grief in general, and grief at the state of the world in particular. Stephen Jenkinson's

Nights of Grief and Mystery[52] are grief-focused arts events that recognize our predicament.

Someone asked me if grief at the state of the world goes away at some point. For some people, it may. For most of us, it comes and goes, but it becomes familiar and not as disruptive as when that grief was new.

Why do I call this chapter "befriending grief"? For me, grief is like a friend, a not-very-polite and sometimes intense friend who likes to visit unannounced. If I pay attention, I hear grief knocking on my heart before it gets worked up, barges through the door, and knocks me over, leaving me bruised. If I let grief in early, it is more gentle. I have learned that it's wise for me to *visit my grief on a schedule* by participating in regular grief circles. About once a month has worked. This rough visitor is a dear friend because a good bout of grief almost always leaves me relieved of fear, guilt, and illusion.

What we grieve

One spring day, Sophie and I hosted a Community Grief Circle online. Seven people gathered on Zoom, from Europe through North America to New Zealand. As part of that Circle, we did the grief ritual described later in this chapter. During that ritual, each of us shared what we were currently grieving while the rest were silent and compassionate witnesses. We did not comment or correct each other. Here is a taste of the griefs we shared that day.

My grief: a friend had asked me the day before if I could afford food. With a job that pays close to minimum wage, she can barely afford food herself. This is ironic, since she works in a grocery store. With her tender heart, she hears people all day long expressing their shock at food prices and their worries about affording enough to eat. Meanwhile, grocery store chains in the U.S. are making obscene profits. Unbridled capitalism is eating itself, and we are in it. My grief contains fury.

Another person was angry at the people who keep doing the things that kill the planet. Why won't they see? Why won't they understand? Why won't they stop? She has found nowhere else to express her grief and anger.

One person remembers when he was swimming next to a whale and her newborn calf, feeling stunned as the great whale's eye focused on him with intelligent awareness. He told us those whales no longer migrate to the site where he met that whale's eye. They no longer have babies there because they cannot get enough food for the journey in the sick and overfished ocean. He thinks of that penetrating eye, and he has no words to offer the whale, just his grief.

Another man grieves for a forest near his home that burned down. In the present climate, it cannot regrow. That forest contained 500-year-old trees. They had survived other fires, but not this last one. He used to take kids on nature tours in that forest; no more. He remembers one child who arrived too scared to enter the forest; it was all he could do to coax the boy to try. By the end of the tour, this kid was beaming with joy and pride. No more; that forest is gone.

Another person grieves for her powerlessness and the painful sense of negation that powerlessness brings. Asylum seekers she cares about are refused. Her conflict negotiation seems not up to the challenges she is asked to face. Young relatives seem adrift. She feels layers upon layers of grief and pain.

Another woman mourns that there are no teachers to guide her through these times. She must try to make her own way.

After we shared these griefs, we processed them with music and poetry. Before we left the Zoom meeting, we blessed each other. After the circle, I felt both a reverence for what was shared and a deep sense of love.

—◦◦◦—

You don't have to analyze or defend your grief, or compare it to others, or assign it a certain value or importance. You may feel lighthearted one day and heartbroken the next. A seemingly trivial loss may feel crushing in the moment. You can practice holding your grief when it arrives: respecting it and giving it some time and attention. And you can practice witnessing another's grief without letting it become yours. I invite you to befriend your grief. This is a way to honor what you love and to learn to live with its loss.

Perspective: Saying yes to sadness

Lisa lives in Victoria, British Columbia. She has been an environmental activist for most of her adult life. At the age of 61, she has witnessed the destruction of many natural places she held dear, and she holds no illusions about the future. She recently had an experience with her doctor that sparked a realization: feeling sad a good part of the time is appropriate to her circumstances, and feeling sad is okay with her. Here she recounts that interaction:

> My family doctor called me to do a mental health check-in. I've been on antidepressants off and on for years, mostly off. But currently, I've decided to just stay on them. They give me a cozy blanket feeling that smooths out some of the rough edges of life but the low dose I'm on doesn't make me happy. My doctor was thinking about changing my meds based on the answers to a test she routinely gives me to check my level of depression and anxiety. My whole life I've failed that test, but I don't actually have a problem with that. A couple of questions about self-harm, I understand, are bad indicators. But the other questions just seem like healthy responses to life as it is these days, and life as it has been since I can remember. I don't think that's a pathology.
>
> I've found support in Joanna Macy's Work That Reconnects, Parker Palmer's Courage & Renewal community, and the Deep Adaptation community. What I love about all of them is they honor the so-called dark emotions. They really make room for them, welcoming them as part of the soul's true expression. The

people in my life who insist on good vibes only or, "Leave your bad vibes at the door," all these kinds of norms, don't stay in my life for long. I attract people who don't need to be happy to feel like their life is good. And I feel like that's me. My mother always says, "All I want for you, Lisa, is to be happy. I just want you to be happy." I've never given her the gift of having a happy daughter. But I also really enjoy my life as it is. There's cultural pressure to do the work to get happy. I'm more interested in working on being courageous and inspiring others to be courageous.

So I had that conversation with my doctor. "OK, you want to try me on these other pills. What goals do you have for those pills that we're not meeting with the pills I'm on now? You must have something in mind about what would be better than what I'm feeling now." She was sort of vague. I think the medical model is vague in general about the function of happiness. What's the point of it? And how do you measure it? So I like to play and poke and be a bit provocative about those norms. I don't think they're that meaningful. I tried her new pills for a week and they made me feel electrified and manic. I hate that feeling, so I'm sticking with the pills I have.

Is Lisa unhappy that she's unhappy?

No, that's the point. Not at all. I won't pathologize it for anybody. I won't say I'm happy about being unhappy. But I'm at ease with it. I think rather than measuring happiness, I'm measuring my life in terms of being courageous and useful. I've noticed I sleep best after I have spent the day being useful and brave. And I don't need happiness for that.

When you are overwhelmed

During the pandemic, I found myself supporting a couple of people who felt overwhelmed. Looking back, they reminded me of some things I said that were helpful. These are all things I have said to myself.

- I can't promise you things will work out. But you are not alone. You matter, and a difficult life can still be a good life.

- Your first job is to work toward keeping yourself as healthy as you can, capable of whatever it is that God (the universe, or your heart) wants you to do.

- Job two is trusting that whatever you don't manage to do, it's okay with God (the universe, or your heart.) So let it be okay with you. Your worth is assured; it is not based on your performance.

- Being humble means remembering that God (the universe, or your heart) doesn't ask you to do the impossible.

- Probably the most important thing to do is to slow down and reconnect with God (the universe, or your heart.) The busier your life, the more you probably need time for reflection/prayer/meditation/nature connection.

Normalizing and simplifying grief

Where Light Can Fall

He asked me what I had for lunch today.
Listing along one side, he rolled one large eye my way.
"I ate the nets," I said.
"I took them in their tangled deep,
"I ate the caught things and every weighted line."
Despair's spaghetti: it weighs heavy.
The light slid along his flank.
The waves rocked me while his
tattered fin slapped at the
perturbed surface

"Do you think you might stay awhile?" I ask,
scanning the horizon for
Telltale signs of trawler's smoke.
"I'd like to," he said, blowing out a flume of air.
"But you have more eating to do, my dear."
A dinner of pollution, I will eat that.
The dessert of population as I spoon each
whipped essence of constant demand.
A cool drink of methane, don't belch. Keep it in.
What does it feel like
to know
we are
killing God?
The seagulls have arrived.
A din, a harridan of judgment.
"You lie! You lie! You lie!" They scream in the salt-bruised air.
They're right. I do.
My hubris is in that I think I can save you.
You push off with a mighty flap and begin your descent
to the heart of the world.
We both know the goal here is to save me from myself.

– Jenna Matlin

Grief is about losing what you love and long for. That is hard enough. In addition, grief can get complicated with all kinds of complicating baggage. Complications often add an excess burden to grief related to ecological and societal collapse. Here are some of the factors that can complicate grief, and weigh us down.

- If our relationship with what or who we lost is complicated, the grief will be complicated.

- "Shoulds" of all kinds: I should have done this. They shouldn't have done that.

- Inward-focused guilt, shame, or a sense of failure if we believe we could have done something different that might have prevented the loss.

- Other-focused blame, resentment, bitterness, or hatred at the people we believe caused the loss or could have prevented it.

- Lack of understanding that what we are experiencing is a loss and that we have the right to grieve it.

- Fear or anxiety if a loss is anticipated.

- A diagnosis of depression or other mental illness, if the grief does not ease as quickly as a medical professional thinks it should or is judged disproportionate to the loss.

- Trauma responses, physical or emotional, if the loss involves violence or the threat of harm, or disrupts your life in a way you don't know how to bear.

- A crisis of meaning, if it seems the world has betrayed you or that there is no purpose to living without what you've lost.

- Being told your grief is inappropriate, out of proportion, or even pathological.

- Feeling alone, unable to name and share your grief with friends or family for whatever reason, or trying to name it and being dismissed.

- Being unable to say goodbye or have closure rituals, as so many people experienced during the pandemic.

Part of the work of grieving is finding ways to let go of this excess baggage and simplify our grief. Just naming the complications packed on top of the grief can help. Hearing that your complicated grief is normal and not unique or pathological may help. Sharing the loss with a compassionate listener or co-sufferer can help immensely. Spiritual traditions offer frameworks and rituals that may simplify grieving. Grief can and should be honored and named publicly. Counseling and spiritual teachings can help to process complications.

—◦◦◦—

I have noticed one type of complicated grief often associated with ecological losses, called *moral injury*.[53] Moral injury happens when a person has violated their conscience or moral compass by witnessing or participating in harm.[54] They bear responsibility for a loss or believe they do, and have not yet been able to forgive themselves. Even if nobody else wants to blame them, they blame themselves. Sometimes a person will cling to self-blame even when clear thinking reveals that they are in no way to blame. They could not prevent the loss, but they persist in the illusion that they could have or should have stopped it. Or simply that they do not deserve to live when another died (human or nonhuman). We can call this survivor guilt. Perhaps they unconsciously chose guilt over an admission of powerlessness to understand or prevent the loss.

Moral injury was first recognized in combat veterans. It is common in survivors and witnesses of all kinds of violence and trauma. It sometimes accounts for what is called burnout. This can happen when a worker is expected to do far more than they reasonably can do. A worker may have to watch people suffer from a lack of attention or care that they are unable to give. They may witness harm they didn't and probably couldn't prevent. This kind of moral injury can be a side effect of collapsing institutions. Moral injury also affects lovers of the nonhuman world who watch a portion of it being destroyed. Living with awareness in these times, we will likely experience some moral injury or survivor guilt. Doing the spiritual work of forgiving ourselves is part of being the eye of the storm.

—◊◊◊—

Complicated grief is normal for people dealing with any kind of collapse. If you experience complicated grief, you are in good company. Seek to simplify your grief, because being weighed down with guilt does not help you serve the life that remains to be nurtured.

Become a tree.
Your limbs no longer move under their own power.
your lower limbs, your roots, are quite stuck
but they are fractal and they move by entangling, they explore.
Your upper limbs dance but not at your command
in the wind.

Mourn the great trees.
Mourn the great trees.
Mourn the passing of the old ones
for the season of trees nears an end.
In the time left, most respectfully,
greet trees great and small.
Touch them.
Lean on them.
Listen at the feet of elder trees;
let them impart their wisdom
while there is still time.
(Acquiring wisdom
is a slow process.
Listen long and well.)
And when they die
do not chop them down.
Let them stand and serve
like standing stones
like obelisks
thrones for raptors
until they finally fall.

– Terry

The Handoff Meditation

I use this meditation when I feel overwhelmed with grief, worry, or responsibilities. Sometimes I know very well what is troubling me. Other times I have to figure out why I am tense, or sleepless, or why my head is spinning. When I am clear on what is troubling me, I may cry and get some relief. But sometimes, I have taken on more than I can carry: worry, responsibilities, guilt, shame, or remorse. I don't know how to put them down. I have learned that it doesn't matter if I think they are mine to carry. If I am overwhelmed, there is no point; I accomplish nothing but making myself exhausted and soul-sick. But I still don't know how to put them down. Here is what I do.

First, I take my time sorting through the burdens. I name them and face them. I may write about them in my journal. Maybe I cry.

And then I imagine they are all in a backpack I have been carrying. I will take off the backpack, and hand it over to someone stronger and more qualified than me. Who? For me, it's always Jesus. Who would always be happy to carry your load for a while?

I can pick that backpack up again any time I wish. Because I have handed it off for a while, I will be rested and renewed, better able to carry it when I pick it up again. Or I may realize that it is not mine to carry at all.

Grief rituals

Longer than I have known almost anything else, I have known that I can survive only by holding gently, like a feather in each palm, the anguish and beauty of being alive.

– Trebbe Johnson[55]

In grief circles I have hosted, words only go so far. We often use the following ritual to honor grief. We also take time "apart together" with music, meditating, dancing, or journaling.

To prepare,

- Gather **smooth stones**, 1-2 inches in size, in a small shallow bowl.

- Fill a **small pitcher**, such as a cream pitcher, with water.

- Find a **deep container,** five inches or more, so that stones added to water in it make a satisfying plunk. If the ritual is for a group, the best container might be a thick solid glass vase so all can see the stones fall.

To do the ritual,

- Name that:
  Stones represent the heaviness of grief.
  Water means the movement of grief, the shifting of feelings and identity that grief creates in us.

- Pour the water from the pitcher into the transparent container.

- Have each person who wishes choose a stone and hold it up. Each person in turn names what they are grieving, in one word or many.

- All gathered say the person's name, and then, "We honor your grief."

- The person drops the stone into the water. Allow a little silence before the next person names their grief.

- If you are doing this ritual alone, give each grief a stone, and be sure to say aloud something like, "I honor this grief." And next time, why not invite a friend?

After the ritual is complete, return the water and stones to the Earth to symbolically complete the cycle of release. When I host this ritual on Zoom, I handle the stones, with an extra camera focused on the water vessel, so we can watch and hear the stones plunk.

—◊—

A friend, we'll call her Amy, experienced the loss of a relationship that felt like a turning point, a change of identity. It brought deep grief. We modified the ritual to include her immersion in the ocean. She gathered her smooth stones. She prepared them in advance, by painting a word on each stone that described something she had lost, like trust, innocence, and dependence.

On the beach at dawn, we gathered with her. She said a few words for each stone and then threw it into the ocean, with her friends saying, "Amy, we honor your grief." Then we guided her into the ocean, with words of invitation to be transformed. She walked into the surf and submerged herself. As she came out, we received her with towels and hugs. She put on a new dress, we each offered her a blessing, and then we ate together.

—◊—

In her book, *Radical Joy for Hard Times*,[56] Trebbe Johnson invites us to honor wounded parts of Earth, near and far. To connect with them and show our love for them, she invites us to immerse ourselves in the place and then create a mandala (a geometric arrangement) of found objects to reflect on and honor that wounded part of Earth. Here are the steps she suggests; a complete guide is on her website.[57]

- Go, alone or with friends, to a wounded place.

- Sit awhile and share your stories about what the place means to you.

- Get to know the place as it is now.

- Share with the others what you discovered.

- Make a simple gift of beauty for the place.

She names this process of witnessing, honoring, and creating "Radical Joy." To talk about "wounded" and "joy" at the same time is a radical move in my culture. Several times I have participated in a Radical Joy ceremony online, with each person using a photo or found object to represent the place being honored, and each time I have found the experience beautiful and moving.

What grief ritual will you make? Who will you invite to help you?

Cultivating joy and gratitude

Even a wounded world is feeding us. Even a wounded world holds us, giving us moments of wonder and joy. I choose joy over despair. Not because I have my head in the sand, but because joy is what the earth gives me daily and I must return the gift.

– Robin Wall Kimmerer[58]

Many people coming to terms with the reality of collapse have a daily gratitude practice. These practices are not meant to cover the real losses of our predicament. They are intended to lift up and celebrate ongoing life in the middle of messiness, hurt, and loss. They are meant to cultivate appreciation, connection, and joy.

Gratitude fills me up. It fills that hole in the soul, that restless emptiness at the root of addictive behaviors that modernity invites us to fill with achievement and accumulation. An interesting thing happens when I slow down and savor the humans and nonhumans that sustain me and bring me pleasure. They seem to expand, and they fill that hole.

A. J. Jacobs took it upon himself to discover and thank everyone he could find who was in some way responsible for helping to produce his morning cup of coffee at a café in New York City.[59] As he worked through all of the baristas, janitors, truckers and ship crews, suppliers, growers, cup manufacturers, and storehouse pest controllers, his list of people to thank ran to a thousand or so. There's a supply chain for you. When I reflect on just a small part of the complex and laborious path my coffee took to travel to me, it becomes a precious jewel that I can sip with awareness, rather than gulp as I dash to my first task of the morning.

If you can't seem to experience joy, be kind to yourself. Perhaps begin with a rote gratitude practice. Try noticing and savoring little pleasant things as they bump into you. If you think you shouldn't experience joy, please uncomplicate your grief and give yourself permission. Your misery does not help you or anyone else.

A life lived fully is a mix of joy and grief, celebration and mourning. The rhythm of those alternating expressions makes for wise and compassionate living. Just as my culture has not allowed public expressions of grief, it has flattened public expressions of joy unless those expressions sell us something. Unfamiliar with exuberant expressions of joy, we might imagine they are out of bounds, shameful, or even dangerous. How sad that our animal bodies have forgotten how to transmit and discharge that emotional energy. We can still learn. Singing, dancing, and drumming can help. Ah, but we are too self-conscious, and not professional enough to do those things, right? Yet bodies want to move, need to move. Churches have choirs or bands for a reason. My favorite exercise is dancing. It is my body making joy.

Summary and reflection

- If you don't need to grieve the state of the world, you aren't paying attention.

- I have discovered that when I am stuck in difficult emotions, beneath those emotions usually lies grief. Grieving what I cannot fix or change brings relief and freedom. That is why grief is a friend.

- Grieving is not respected or understood in our culture. *We have not been taught how to grieve.* It's time to learn.

- Lisa accepts her sadness as a normal part of living in these times.

- Grief can be complicated in various ways, especially grief for the nonhuman world. If we can simplify it, it will weigh on us less.

- Living with an awareness of ecological ruin in these times, we will likely experience some moral injury or survivor guilt. *Doing the spiritual work of forgiving ourselves is part of being the eye of the storm.*

- "The handoff" is a simple tool for taking a break from what is burdening you.

- Rituals help express and release griefs of all kinds. What grief ritual will you make?

- Grief can guide us to what we truly value. We can find joy and gratitude on the other side of grief, or even in the midst of it. Gratitude lists, and practices that cultivate joy, are powerful tools for living well in hard times.

CHAPTER SIX

Belonging and reverence

In this chapter, we shift from coping with emotions to cultivating meaning. When foundations shake, that which is not deeply rooted tumbles over. Modernity centers on the individual. A modern person's identity often has only a shallow rooting in vulnerable things like their immediate family, job, or economic or social status. Being the eye of the storm requires being rooted in a deeper and broader sense of belonging and identity. It means knowing and trusting in life-giving realities, seen or unseen, that are our source and home.

Deep belonging. Belonging to something larger than yourself is a strategy for finding meaning and stability.

Perspective: Invitation to Earth Listening. Emma Mary has found a way to help others to hear Earth's wisdom and love.

So many precious jewels. An invitation to find wonder and beauty in everyday life through a simple "Precious Jewel Meditation."

Perspective: A garden gift. L. Z. discovered a profound connection while planting fruit trees.

The Honorable Harvest. Robin Wall Kimmerer invites us to be in a reciprocal and respectful relationship with that which sustains us.

Perspective: Reconnecting with nature. Jay broke out of her modern scientific approach to experience the natural world more deeply. Now she offers groups to help others do the same.

No simple acts. Finding meaning and value in everyday activities.

Deep belonging

You are an island of forest in the desert canyon,
your fingers peninsulas stretched over empty space below.

You offer my feet shade, soft earth, stones, pine cones,
waving clumps of grass, and fertile dust.

You are dressed in firs, some stretched tall
like Dr. Seuss Christmas trees.

You are rimmed in aspen, revealing the path of wind,
remembering spring,
with white bark and pale shimmering green leaves.

You reveal lone sentinels, stout poles, some greening in power,
others black and white on an empty slope,
remnants of a society of conifers cut and carted away.

You speak as wind, as robin, as cicada, as buzzing fly.

You dance as swirling aspen leaf, as wavering butterfly,
as bowing head of grass.

Your layered cliffs expose hidden stories of fateful acts,
buried beneath layers of silence.

You do not allow me to hide.

Your demanding gusts bare my soul till, longing for sleep,
I begin to write the hardest chapter.

You hold death, memorial stands of black pine poles.

You hold birth, windflower buds
and bright tips of new evergreen, like tiny bouquets.

You hold secrets, verdant dripping springs
hidden under rock roofs.

You hold change: one hillside in aspen, the next in fir,
another a graveyard of pine poles.

These things tell me that I am part of something
much older and vaster than I can know.

– Terry

Belonging is a foundational human need that drives much of our behavior. People will go to great lengths to try to satisfy their need for belonging, and with it, identity and security. Craving those basic needs, people can fall prey to authoritarian rulers, scapegoating, or sectarian violence. This is why a direct personal experience of deep belonging that does not depend on human leaders is so valuable in hard times. Such experiences help us resist people claiming to know and control our reality.

Here is my personal story of deep belonging. How you find belonging may be very different. Do what works for you.

I experienced the Roman Catholic religion of my childhood in two different ways. The most obvious way was the ethical and ritual observances that my parents expected of me. We recited beliefs from memory during worship but never discussed them, not even in my religious grade school. I wonder if anybody believed them. But another recurrent and subversive thread said, "If you take this religion seriously and pledge yourself to it, it will transform you. You will meet the Holy and answer to it directly. You could become a living witness to its reality, its servant in the world, and you would then thumb your nose at the world's values in favor of something far more precious."

This thread showed up in three places. First, in the gospel stories I heard each week in Sunday Mass. The interpretations may have been conventional, but the gospel stories of Jesus always carry invitations to mystery and countercultural values.

Second, in the adventures of several of my elder relatives who had "got the call" to become priests or nuns.

Finally, my grandmother's bookshelf held a handful of children's books about the lives of the saints. The plot of each book was the same. A young person encounters the sacred, then must disobey their parents (with much drama) to follow the call they have been given.

I was close to thirty and married before I had my own encounter with the sacred. I didn't disobey my parents, but I did thoroughly shock my atheist husband. That encounter with the sacred has shaped my life.

Because I had already rejected the dogma I inherited, I never thought of myself as having strong faith. Imagine my surprise when I took a "spiritual gifts inventory" years later and got the highest score on the gift of faith. What? Why? But on reflection, it made sense. That visionary experience of a loving and accepting Jesus still feels so real that I trust it entirely. I do not need belief. To quote Carl Jung, "I don't believe, I know."[60]

Mysticism, firsthand experience of the sacred, is well known as a threat to religious authority. It is also a wellspring of life and meaning for many, as it is for me. Seeking a foundation for my mysticism, I discovered Matthew Fox's Creation Spirituality.[61] Fox was a Catholic priest who dove deep into the bible and the Christian traditions to lift up threads of mysticism, social justice, and Earth-based spirituality. He helped me to claim the nonhuman world as sacred. His Earth-affirming theology was not tolerated by Catholic leaders, so now he is an Episcopal priest.

Spending time with Fox and friends, I discovered a variety of agreeable practices to feed my hunger for experiences of the sacred, including Dances of Universal Peace and Harner Core Shamanism. Journeying, Harner's foundational tool, invites relationships with the nonhuman world. His neo-shamanic system has allowed me to seek the counsel and power of compassionate helping spirits. My repeated queries to Jesus have been met with repeated reassurances that he is okay with me seeking help and power from these other spirits; they are on the same team. I know I am never alone or without resources. Earth Listening with Emma Mary Gathergood and other participants has been a recent source of deep connection for me. I have a lifeline and a source of strength through my connection with the sacred as I understand it. I belong to something ancient, powerful, wise, and loving. Nothing can take that away from me.

I am not unique. We are all connected. Our Earth home is cloaked in a radiant web of life. What one does affects all. And, in my culture at least, we forgot. We forgot we are all connected. Now we get to relearn in a hard, hard school. Please, find your deep belonging. Get it firsthand, or borrow it from someone with a faith or experience of the sacred (of wonder, of what matters most, of what endures... don't get hung up on the language) that seems agreeable to you.

If you need deep belonging, consider repurposing the tradition you were brought up in, if any. Every tradition has its glaring hypocrisies (well, every tradition with a paper trail does.) And every tradition, entered deeply, will challenge you to stretch further than you are comfortable going. If you were brought up in a certain tradition, that is the one you know best, so practicing it saves time. You already see its hypocrisies; that's convenient. And that tradition is the one whose demands you are well aware of, so you can spend less time fooling yourself that any spiritual path is free of challenges. These are the arguments that Chogyam Trungpa made to me in his book *Cutting Through Spiritual Materialism*.[62] He kept me Christian.

Religion is not required for deep belonging. Do what works for you. Contemplating the nonhuman world, especially at very large and very small scales, is a practice many nontheists use. This kind of contemplation can create a state of wonder and trust. It can be a comfort when ecosystems and human systems and many lives are being lost. You can reflect on the following:

- Being made of the dust of stars that died billions of years ago.

- Hosting multitudes of ever-renewing cells, and cells-within-cells, in one human package.

- Wondering what the rest of the universe is up to as we struggle with our predicament.

- The stories trees have to tell from their long lives.
- The stories stars have to tell from their longer lives.
- The stories insects have to tell from their short lives.

And you are a part of so many other wonders. Contemplate the ones that bring you deep belonging. Wider Embraces[63] is a style of group meditation that explores our belonging at different scales. Many other forms of meditation and contemplation can also cultivate a sense of deep belonging. If you have access, physically immersing yourself in the nonhuman world can be another way of experiencing deep belonging.

Your deep belonging will get tested when hard times come. Facing the storm means facing core questions of identity and meaning. What is required to make a good life? Who are we, really? Who are we if we lose familiar stories, identities, our home or livelihood, or our relationships? What kind of life is worth living when times are hard? A sense of self that can withstand the disruptions we face is found in wider and deeper identities than modernity has to offer. A secure sense of belonging to a family, community, place, culture, or spiritual tradition, is an anchor in the storm of climate and social chaos.

In the worship of the Great Mother,
In the teachings of the Son,
In the songs of the priests and prophets,
They forgot

That the Great Mother walked barefoot
To the well each morning,
Tended chickens, grew herbs,
And minded her sisters' children.
That the Son worked wood, stoked fires,
Knew the names of all the trees,
Judged them sound for working or suited for firewood.
That the priests and prophets sang

Of hills dancing for joy, stars portending wonders,
Long treks through dry lands,
And the delight discovery of a cool spring.
They forgot.

They exiled themselves from the Garden
And scarred the Earth
To build great stone halls grasping for the heavens,
for they had judged the Earth
And her raiment and her friends
A distraction from striving, striving, striving.
Striving to obey
Striving to belong
Striving to buy safety,
Striving to avoid the fate of those judged not worthy.

In the worship of the Great Mother,
In the teachings of the Son,
In the songs of the priests and prophets,
I remember

Dusty feet, gently shorn of worn sandals,
Dipped in cool spring water.
Flatbread flecked black from resting on glowing coals,
Messy births, sweaty farming, lusty singing,
Dancing under the stars.
I treasure

The earthen trail trod hard, the earthen cushion under oak shade.
The sacred canopy of vast evergreens, the carpet of cool duff.
The tree rings telling their tales of times beyond time.
The tiny creatures sheltering in the crumble
beneath a vast fallen giant
Who are also the Mother's beloved children.
The dripping fruit that brings a smile.

– Terry

Perspective: Invitation to Earth Listening

Emma Mary lives near her beloved Malvern Hills in west-central England. As a therapist and a holistic life coach, she listens well to people. Recently, she started intentionally listening to the Earth regularly, and helping others do the same. She has been hosting Earth Listening circles in the Deep Adaptation Forum since 2020. She repeatedly tells participants, "We say we love the Earth, but remember the Earth loves us too and wants to be in relationship with us."

In our Earth Listening group, Emma Mary invites participants to visualize a beloved spot in nature, preferably a place close to home. She offers a brief reflection on a topic for the day and then invites participants to listen to the Earth for a five-minute silent meditation. If they choose, people share briefly after the meditation. Reflections run the gamut from self-aware insights to images and visions to conversations with nonhuman entities. Whatever the form of the messages they share, participants almost always receive multifaceted wisdom, right for each person, as well as frequent humor and tangible care.

Emma Mary's current practice of Earth Listening arose from a stern talking-to by the angry Welsh sea. She had rented an Airbnb cottage on the seacoast, in the little town of Tywyn, for a holiday with her partner. Earlier in the holiday, she had thrown her back out. Eventually, all she was good for was sitting on the bed with the balcony doors wide open and gazing at the sea. She did that for hours each day.

Near the end of the visit, she suddenly sensed that more was happening at the shoreline than the high tides of the March equinox. The sea was wild, and it started talking to her. It told her how angry it was, how incredibly angry. "Humans purportedly love the sea. So what do they do? They go and build all these horrible holiday homes and hotels along the seafront and just pollute it. They use it for sports, but there's no connection. How can they do these things, burying nuclear waste and pollution and throwing all their garbage into the sea? Don't they know what the sea is?" She felt a sense of incomprehension, "What are they doing?"

All this happened during a 20 or 30-minute meditation. The sea was demanding that she listen. When she finally went downstairs, looking quite wild herself, she gave her partner a bit of a fright. He would have been more disturbed if she had told him what she had just experienced, so instead, she wrote in her journal what the sea had told her.

Two weeks later, she decided to check out the Deep Adaptation Forum. Just as the COVID lockdown started, she found herself in a Zoom breakout room with two lovely people. She got up the nerve to tell them about listening to the sea; she hadn't told anyone else about her experience. Instead of dismissing her or questioning her sanity, they both said, "Yeah, we like to listen to the Earth. She speaks to us too." Emma Mary was delighted to meet kindred spirits. The three exchanged emails and set up a little Zoom meeting, where they could be strange and different together. A few more friends joined them, and that group met weekly for over half a year. They took turns leading meditations on listening to the Earth.

By fall, Emma Mary realized that Earth Listening needed to reach more people. With tech help from one of her original listening partners, Emma Mary made the group open to participants of the Deep Adaptation Forum. Since they started, the group has met every Friday without fail, except during the weeks of Christmas. One of her participants now holds twice-monthly sessions for people in Australian and Asian time zones, and others are now offering sessions in Lebanon and Germany. For information, see earthlistening.net.

On Sunday afternoons, she asks Mother Earth what the topic will be for the coming week. To be precise, she asks a favorite grove of trees on her beloved Malvern hills, visible through her window. While her coaching and therapy backgrounds inform her reflections that start the meditation, the messages people receive take those meditations further than she would ever have imagined. All is not sweetness and light; sometimes the meditations are on anger, or on loss and mourning.

Many people have passed through the groups now. Some are regulars, some occasional. Emma Mary's approach is decidedly too "out there" for some people. Or is it? One curious person made it their task to visit every meeting listed on the Deep Adaptation Forum calendar for a certain week, or they would never have joined such a "woo-woo" group. That person had a transformative experience during Earth Listening.

Emma Mary is continually learning, delighting in what nature discloses and what participants in her groups discover. She likes to remind us of this quote from Robin Wall Kimmerer: "Knowing that you love the earth changes you, activates you to defend and protect and celebrate. But when you feel that the earth loves you in return, that feeling transforms the relationship from a one-way street into a sacred bond."[64]

So many precious jewels

In my vision:

> *I am swimming around a reef. I am led by my teacher to dive into a deep coral cave. Somehow, the cave is dry. The walls of this cave are encrusted with glittering faceted jewels. My teacher begins plucking these jewels from the walls of the cave. Blue, pink, gold, green, violet.... so beautiful. She collects them from the walls, then stretches out her cupped hands and tells me to put these jewels in my pockets. "Look for the treasures, and treasure them." I am confused. I ask, "What is my work?" She smiles and sighs, "This is not work. Honoring the beautiful, lifting up the beautiful. Caring for the beautiful. And by the way, it's all beautiful." As she says this, my heart expands.*

Industrial consumer society has a motto: more. Its purpose is to make more, use more, and hoard more. We are promised more. Many of us expect more. Demand more. We need more to have enough, to be enough, but it is never enough for long. Addiction, bitterness, debt, depression, or burnout can all come from chasing more.

Beneath the cry for more is a vast emptiness. No matter how much money or how many possessions we collect, this hole remains empty and gaping. It is easy to believe that more is needed to fill this hole. But the hole is unfillable. It tells us we never have enough. It tells us we never are enough.

This hole also awaits those of us who want to be of service. We expect more of ourselves than we know how to deliver. We want to please everyone, or to accomplish something more meaningful, more effective, or more lasting than we can achieve. It is never enough. We are never enough.

What is an antidote to the emptiness that demands more? For me, it is the perspective that the things that make up our ordinary lives are actually precious jewels. We too, are precious jewels. Flawed, yes, but also radiant. When I see from the perspective that everything is a precious jewel, sparkling and wonderful, I am already incredibly rich and full of good things. I am immersed in a wondrous wholeness that fills the hole. Familiarity can prevent me from seeing each little thing surrounding me as a precious treasure, but so it is from this perspective.

Precious Jewel Meditation

My current favorite meditation, based on the vision above, is to observe a proper reverence for the treasures surrounding me. I simply take a look around and savor. As I write, I don't even have to move my eyes to see the computer that makes this book possible, the wrist brace that allows me to write, the glasses that allow me to see, the 30-year-old cherrywood slab table that holds so many memories, the Pomodoro timer that saves my back from damage, the table runner naturally dyed and intricately handwoven in Peru, the little notepad that serves as my auxiliary brain, a tray of drying sage leaves that scent my day, and the fuzzy-leafed African Violet that practically blooms in the dark. Glancing up to the window, I see my garden friends in spring profusion. Ah… I have enough. I have far too much. I cannot properly thank and appreciate each person, place, and thing. Savoring the precious jewels surrounding me is a daily practice for me. With it, I am filled.

Perspective: a garden gift

L.Z. works at a community garden. She does as much as she can to keep it going, to make places for the birds and other small creatures to have a habitat. Her region gets no rain for six to eight months of the year, so small things like ensuring the bird bath has water can make a big difference. She wants to share this experience in hopes it will encourage someone else. Here is her story:

> The community garden was gifted with 30 dwarf fruit trees. I convinced the other members to plant some of them outside the fence along a path that connects the street with a schoolyard. I was on my knees, planting a tree, wondering if the tree would even live long enough to bear any fruit. I had my hands in the dirt, and suddenly, I felt something inside my skull, like a switch going off. Or on. Maybe it was a stroke. It's hard to describe, but I felt like my skull was full of warm, yellow light. It was very comforting. And I was filled with love for that tree, and all the trees, and the garden, and all the plants and critters in it. I remembered how just that morning, I had filled up the bird bath, and all the birds came swooping down to splash and bathe and drink. And I remembered the words a very wise person had spoken to me a while back when I was asking them how they could remain so sanguine in the face of extinction. That wise one said, "It's not dead here." Truly, the garden was not dead. It was teeming with life. And just at that moment, inside my head, She spoke to me. Her. The Mother. Gaia herself. She spoke to me… ME! I am no one, and she spoke to me. She said, "Don't worry about me. I'll be okay. I just need a little help."

It was a profound experience, and I have not been the same since. I have physical problems, and I am not young, but I found that I was able to do a lot more than I could before. I hauled wheelbarrows full of wood chips out to those trees.

So, I want you to get out there and create some habitat. Plant. Make a guerrilla garden.[65] Put out a bird bath and keep it clean and filled. DO something. And expect to be surprised.

The Honorable Harvest

Action on behalf of life transforms. Because the relationship between self and the world is reciprocal, it is not a question of first getting enlightened or saved and then acting. As we work to heal the earth, the earth heals us.

— Robin Wall Kimmerer

We must take from the earth in order to live. And we can do so respectfully and considerately. Well, at least in theory. We can certainly move toward that respect and consideration. It's worth the effort. For guidance, author and scientist Robin Wall Kimmerer, a member of the Citizen Potawatomi Nation, offers the idea of the Honorable Harvest in her book *Braiding Sweetgrass*.

Collectively, the Indigenous canon of principles and practices that govern the exchange of life for life is known as the Honorable Harvest. They are rules of sorts that govern our taking, shape our relationships with the natural world, and rein in our tendency to consume—that the world might be as rich for the seventh generation as it is for our own.[66]

People who have lived in a place for generations have usually learned how to nurture the land as it nurtures them. We are blind to much of the destruction behind our daily lives, but we can begin to learn and change. We need not claim to be Indigenous; we can simply honor our Earth home by learning the real cost of what we consume, and we can honor the particular places that shelter and feed us by developing relationships with them and tending to their well-being.

The Honorable Harvest is an invitation to consider every entity that contributes to our daily life *as a gift*, to use respectfully, with an eye to the care and thriving of that entity. It is part of a larger theme of reciprocity as the basis of respectful human living with the non-human world about which Kimmerer writes so eloquently. Here I reproduce her suggestions for the Honorable Harvest, based on traditional teachings.

- Know the ways of the ones who take care of you, so that you may take care of them.
- Introduce yourself. Be accountable as the one who comes asking for life.
- Ask permission before taking. Abide by the answer.
- Never take the first. Never take the last.
- Take only what you need.
- Take only that which is given.
- Never take more than half. Leave some for others.
- Harvest in a way that minimizes harm.
- Use it respectfully. Never waste what you have taken.
- Share.
- Give thanks for what you have been given.
- Give a gift, in reciprocity for what you have taken.
- Sustain the ones who sustain you and the earth will last forever.

I gulp when I hear that last word. Forever. Have we already shut down that possibility? But I don't need to know how far off the path I have strayed to start searching for my way back.

Take what is given? How do we know what our Earth home gives freely? If possible, we can ask traditional knowledge keepers what is appropriate to take, when, where, and in what quantities. Whether or not that is possible, try having an imaginary conversation with the nonhuman entity you want to take, or just sensing its intention. You might be surprised by what you discover. In Potawatomi thinking, as in most Indigenous thinking, those entities are persons, deserving to be respected and heard.

On an everyday basis, we can begin to see everything we consume or discard as a gift of the Earth, and consider its true value. We can educate ourselves about where our food comes from and who grows it, where our garbage and our plastic end up ("recycled" is not a place and often a lie), and we can take the answers to heart.

The Honorable Harvest is traditional wisdom for foraging sustainably in areas held in common, the areas industrial people incorrectly label wildlands. It is not a complete guide for people in densely populated areas; those lands would be quickly denuded if each harvester took half of an entity. And please respect U.S. State and National Park rules that typically forbid removing anything from the park. This is a wise policy for the countless visitors who could strip a place bare if each took "just one."

The Honorable Harvest is not a guide that fits comfortably into my grocery shopping routine. Instead, it is a powerful invitation to be in a reverent and reciprocal relationship with all that sustains me. Bringing this sense of responsibility and care to our lives will be an act of healing. How can I practice the Honorable Harvest as an alternative to the instant, delivered, plastic-wrapped, disposable, paid-for, taken-for-granted, stuff that makes up my unsustainable living in the suburbs? I'm not sure, but I intend to have fun experimenting.

Perspective: Reconnecting with nature

Jay lives in Australia. During the pandemic, she relocated from Canberra in the south to the Sunshine Coast, the eastern coast north of Brisbane, to be with her partner and closer to family. She told me about how she started facilitating nature connection workshops and guiding "forest therapy" walks.

Shortly before the pandemic reached Australia, Jay decided to participate in an intensive nature connection retreat. Before she could allow herself to attend, she felt compelled to read the scientific research on the effects of exposure to nature on human health. Those effects are real.[67] She didn't need to convince herself. She needed a claim to legitimacy if somebody asked her why she was doing this "nature therapy thing."

That intensive retreat in nature allowed Jay to have emotions that she normally wasn't able to "meet herself" to have. Nature therapy feels true and nurturing to her. She didn't imagine that she would quit her job and be guiding walks a year later. She had thought of the retreat as a little holiday.

Then the COVID lockdown happened. This allowed her plenty of alone time to do all the suggested follow-up activities from the retreat. She learned about the geology and history of her local woodlands, including some history of the people indigenous to her area. She learned to create rituals. These different ways of relating to place all fit together, and she felt that she was learning something transformative.

Her connection to the more-than-human world widened and deepened, and she began to wonder if she could lead people on the same journey she had taken. She has made a start. She is leading small groups into the local forests for meditative walks and circle-style reflections. Preparing for these groups leads her to spend more time, more intentionally, with the land she loves. This was a delightful discovery: bringing others to nature is deepening her own practice.

She wants to foster in people a recognition of their nature relationships, for those experiencing grief, eco-anxiety, and collapse-anxiety, and those who right now just feel harried by the pace of daily life. Slowing down, honoring a place by getting to know it, witnessing interconnection and nonhuman lives, learning and imagining traditional ways of relating to the Earth: she believes all these are small ways to break the cycle of mindless exploitation that is hastening our collapse.

No simple acts

Know today all you cannot repair,
So that you are free for all you can.
Let a hundred things be left undone,
For you to do the one that matters most.
Maybe you will cross a stream and look for what lives in it.
Maybe you will cross the expanse of your spirit
And look for what lives in it.
Maybe you will be there when the cardinal comes to drink,
Or the eyes of someone you love turn to your face.
Maybe you will grow something or harvest something.
Maybe you will make a stitch or let a prayer rise.
These are no simple acts.
This is your foundation.

— Laura Martin[68]

We don't need heroic acts to find meaning in our lives. We can rediscover the wonder of creating the things of everyday life and the satisfaction of tending and befriending: these things are honorable and crucial work.

My family and my church taught me that a vocation provides identity and purpose. It was the work I would do, for which I would be educated, from which I would gain a livelihood, meaning, and satisfaction. My dad liked to say, "Go to college, so you don't have to dig ditches." As a first-generation college graduate and a physicist, he preferred his vocation to his own father's long hours with low pay driving a cab. College and a salaried vocation have been one ticket for participation in the middle- to high rungs of the ladder of industrial consumer society. The ladder of worth claims that higher-paid work and work requiring more education are valuable. Low-paid or unpaid work, so-called unskilled work, is considered low value. We learned otherwise during the pandemic when many higher-paid workers could stay home and do whatever it was they did or didn't do. Meanwhile, many low-paid workers were recognized as essential and faced daily exposure to COVID, often with no paid health care or sick leave in the U.S.

Industrial consumer society has given (some of) us the means for living so easily that we have forgotten that the "simple acts" of sustaining our lives and those of our family are honorable parts of most peoples' vocation. Specialization and money transactions have removed us from the wonder and labor of growing food and making homes, clothes, and tools for living. This is the stuff of life that we have outsourced and take for granted, and that outsourcing is not sustainable. We have forgotten to appreciate and celebrate these simple acts for life. If, or when, we lose industrial consumer society, we will relearn this appreciation. Why wait?

Vocation comes from the Latin word *vocare*, to call or name. To be called (by God, the universe, your own heart) to particular work for a time is a wonderful subset of our foundational call. That foundational call is to be human in our unique way, one of Earth's precious jewels, shining forth to enrich our interdependent Earth home. In a time when the stories of industrial consumer society are failing to guide us, this call is renewed every morning, inviting us to take small steps to shape the unique life that is a precious gift to us. Living with an awareness of the fragility and wonder of what we have, we can remember that daily living is not trivial. There are no simple acts.

Summary and reflection

- Being the eye of the storm means being rooted in a deeper and broader sense of belonging than your individual identity. Your sense of belonging will get tested when hard times come. My deep sense of belonging comes in part from my religious experience. Religion is not required for deep belonging. Do what works for you.

- Emma Mary listens to the Earth and helps others to listen too. Her practice of Earth Listening meets that need for belonging and identity, and offers wisdom from the Earth as well.

- Savoring each person, place, or thing I encounter and realizing what a precious jewel each one is, I am filled with wonder and gratitude. This is one way I overcome the unending need for "more" that industrial consumer society cultivates.

- L.Z. shares an encounter with the more-than-human in her community garden. It energized her, and she has not been the same since.

- The Honorable Harvest invites us to shift our way of interacting with the nonhuman world. The goal is a reciprocal relationship, giving as we take.

- Jay shares her journey of reconnecting with nature and helping others reconnect too.

- We can rediscover the wonder of creating the things of everyday life, and the satisfaction of tending and befriending. These things are honorable and crucial work. *There are no simple acts.*

CHAPTER SEVEN

Resigning from the rat race

Empowered with stories for making meaning and tools for settling emotions, we are ready to experiment with ways of living differently from the mold of industrial consumer society. In this chapter, you will find reflections and experiences that challenge my culture's ideas of achievement and busyness: less doing, more being.

Slowing down and paying attention. A countercultural invitation to work at the speed of relationship.

Not going to pieces. Everybody wants a piece of me. I am learning to say no.

What really matters. An invitation to get clear on your values and tell others about them.

Perspective: Stop striving. Josie talks about stepping away from a job that was slowly destroying her.

Step off the ladder: A plea for earthiness. Climbing the precarious ladder of worth is modernity's way of keeping us anxious and unsatisfied.

Preserve lessons. Stewarding a plot of land is teaching me to slow down and tend relationships, human and nonhuman.

Perspective: Tom's reflections. Tom values being present to the gifts of life more than doing.

Slowing down and paying attention

The times are urgent. Let us slow down.

– Bayo Akomolafe

Being in the mapless territory that is the decline and collapse we face might seem a cause for urgency. That's panic talking, and panic does not help us find our way. Since we don't know where we're going, going faster won't help get us there. Best, then, to take a deep breath and slow down.

We can only understand what we take the time to learn, know, and be in relationship with. To be in relationship beyond the human world is to be always slipping into a state of wonder and reverence. Cultivating reverence for the human and nonhuman world slows us down. The good gardener is always poking about, seeing how her leafy friends are faring on this day. Each ingredient has a story of the journey it took to our kitchen, and each ingredient is owed a word of thanks. We would be wise to work at the speed of relationship.

Staying busy is a great way to avoid seeing what is in front of our eyes. It takes time to truly see. I am a most annoying hiking companion. I have to greet the monkeyflower blooms I haven't seen since the year before, examine animal tracks, watch the clouds, and see if the pearly everlasting seeds have scattered. What's the point of visiting my friends out there on the trail if I don't take the time to greet them?

The human world, too, takes time to understand. Where are the levers of power in my city? Where does the recycling go after it's sorted? How do children learn our values? What are our values, truly? The ones we speak, or the ones we have been living unknowingly? Answering these questions cannot be done at speed.

I have learned in my volunteer and activism work that relationships take time, and relationships are what make non-paid groups work. So if I want to have an impact, I will slow down. I will work at the speed of relationship, and not faster.

I am known as a gardener among my friends and consulted for advice. This reveals the sorry state of gardening in the circles I travel, but I do what I can to help. A colleague of my husband consulted me about her dying houseplants, and I honestly can't recall what I told her. I saw her a few months later, and she said, "Your advice worked!"

I had to ask, "What advice?"

"Pay attention. You told me to pay attention."

Not going to pieces

Don't run around and panic. Don't light your hair on fire. Don't go out and see what other ten more things you can do, or how many articles you can forward… just stop and get really, really quiet and touch down into the Earth and really listen and see what comes up into your heart.

– Dahr Jamail[69]

Everybody wants a piece of me. In 2018 I made the mistake of donating monthly to a political campaign fund. They promised that they would divide my donation among senatorial candidates of my political party for each of the fifty states. They stopped charging me after a couple of months when my credit card number changed. But from then on, I have received email requests for donations from dozens of candidates, covering most of those fifty states. I do my best to unsubscribe, but the deluge of emails keeps coming.

In the same way, invitations from activists and volunteer groups in my local area often leave me feeling pulled in fifty different directions. In my community, I meet the same activists helping each other on about a dozen causes and organizations. They look seriously overextended. I love these people. They have such big hearts. I want to spend time with them. And I am learning to admit that there is no more of me to give.

I want to show up fully when I've promised, or else not promise. To do that, I have learned, I need free capacity. That way I will likely be available when a group I am already supporting asks for a short-term extra. I want to be very clear about what is mine to do, and give myself permission not to do what is not mine to do, and not to feel guilty about it. This is hard when people are hurting, and the list of emergencies keeps growing. It's a work in progress. But I am determined not to go to pieces.

How to say no? Here are some of the strategies I am trying.

- Affirm the value of the request. Somebody wants to relieve suffering or to share some fun, and asks me for help. I can tell them their cause is worthy.

- I can say, "Let me think about that," even if I want to say yes right away.

- If I want to show care for the person, I can ask them a few questions about their project and listen with empathy and curiosity. I can also ask about them as a person, not just a project.

- If it feels like I am the only one who can help, I try to remember that is almost certainly not true.

- I can remember the last time I signed up for something I regretted.

- Admit my limits. "I'm so sorry. I don't have the capacity for that."

- I set some limits around phones. "I don't give money to phone solicitations, so I won't waste your time." Then I hang up. I don't answer calls when I'm in the middle of something. And I don't spend time talking with those who don't show me respect or courtesy.

- I set some limits around email and texts. I take time to unsubscribe from lists I don't want. I turn off email and text alerts.

What needs doing and what you can or should do are two entirely different things. As hard as it is, don't think you are responsible (in other words, that you must respond) to all pleas for help. Please don't go to pieces.

What really matters

Life is too short and too precious to waste it living out someone else's values. We must find our own.

— John Norman

What really matters? How do we want to do and be in the world? The human response in practice to these profound questions in most times and places is, "What I always have done and been. What my family, friends, neighbors, and trend leaders tell me." We run on autopilot. This approach is particularly dangerous when social media manipulates our sense of what is unacceptable and what is normal and right. Or left.

What you value is what takes your attention. Sobering thought. Now that you know, you can seek strategies to spend your time in ways that honor your aspirational values rather than those values your culture wrapped around you like chains. The end of "business as usual" is a great time to examine our values. Once we step away from denial of the troubles we face, we are in a different landscape. The values others take for granted may look irrelevant or even suicidal.

We may practice the values of our sick system unreflectively even if we would deny their hold on us. Take note of "default" values you habitually act out that don't serve you. Challenge yourself to challenge them. You can do this simply by observing how you spend your time.

Hard times allow us to discover what we truly value. It takes time and reflection to rebuild a sense of what does matter, and how our values lead us to do and be in the world. Be patient, and recognize that we are all under construction. Expect your values to continue to shift as you envision ways of being other than industrial consumer society. Give yourself grace when you see the gap between your values and your practice. If you are like me, you are entangled in a system whose life-denying values you don't know how to escape. Living in two worlds is disorienting, yet I don't know how to do otherwise.

I urge you to practice *naming* the values you want to practice. Find an "elevator speech" for them. It is easier to do a thing when the thing has a name. In the Deep Adaptation Forum, our purpose statement is *enabling and embodying loving responses to our predicament.* People light up when they hear that for the first time or the thirtieth. The Permaculture community has the values: *earth care, people care, fair share.* If your values don't sound alluring (at least to you), you either haven't named them yet, or you haven't quite found them yet.

From the values and purpose you claim and your particular history, location, and passions, you will make a life despite hard times. You will explore ways to be and things to do that serve life in your unique situation. If you don't know what that looks like, spend time exploring. It might turn out very different from what you can initially imagine. Or it may be what you were already doing, valued anew, for different reasons. Either way, finding new or renewed values may take time and experimentation, and the support of others.

Don't assume you must be a shining example of the purpose you name and seek to practice. Your work in progress, that a critical person might label hypocrisy, is normal. You are human, not perfect.

Brag on your values, even if you feel sheepish about how you embody them. *In hard times, people desperately need values and purpose. They make life worth living.* Somebody may take your values and find a path of action different from yours. That's to be expected. A good set of values isn't a map; it's a way of traveling.

Step off the ladder: A plea for earthiness

For many people, living life means fighting for a place on a precarious ladder instead of having two feet firmly on the ground. Are you familiar with this ladder? It is the ladder of worth. If you are up high on that ladder, you feel good. You are really somebody, looking down on all those other people. You are better than, but that just feels like being good enough. You have arrived. But you are not secure. One day you may slip and bump, bump, bump, down the rungs you go. That slip makes you less than. When you land down in the sub-basement rungs of the ladder, you feel worthless. You feel shame.

What sends you up and down that ladder of worth? We each have our personal list. I ascend the ladder because I'm so smart. I have a Ph. D. But do one dumb thing and bump, bump, bump, down I go. Maybe your ladder is about having enough money or an important job. Perhaps it's about getting someone's good opinion. Ladders work in the same way, whatever their rungs are labeled. You can find so many reasons to fight your way up the ladder. And so many ways to bruise your behind slipping down that same ladder. You can climb a ladder because you are so in touch with the sacred, so tuned in, so filled with the Spirit. You are ascended. Then, you go through a spiritual dry patch or stick your foot in your mouth, and down you go. Bump, bump, bump.

Here's a secret. You can step off that ladder. You can stop comparing yourself to anybody else and settle into the place you truly belong, both feet firmly planted on our Earth home, resting securely in the great web of being. You can stop trying to earn your place and stop judging yourself and everybody else. You can take your feet off those rickety slippery rungs and plant them on terra firma.

With both feet on solid ground, you can relax. You have nowhere to fall and nowhere to climb. You are secure, and you belong. You are shoulder to shoulder with everybody else, not above or below. There is no pinnacle to which you must ascend to receive your value, no shortage of worth to fight over. You are part of a whole that is all sacred, all worthy. In this view of the world, we never were less than. And we never needed to be better than. We belong. We are accepted, valued, and loved as one small and precious part of a sacred whole. Secure in that reality, we are grounded and steady when the storm comes.

Picture two people on a ladder trying to hug or make eye contact. It doesn't work. A ladder is a lonely place to be. We can stand shoulder to shoulder when we step off the ladder onto solid ground. We can see each other eye to eye. We can hug.

"What about all those other people, still on the ladder judging me?" Yes, they're still judging us. That's their business, not ours. We need to find people with both feet on the ground. These people will support, guide, and cheer us instead of judging us.

So come and discover this wide-open place where you can relax. Here you can be honest about yourself and not be shamed. You can learn and make mistakes, and not be a mistake. Quit climbing an infernal ladder to some imaginary destination or sliding down into a pit. You can spend your energy loving, serving, and enjoying, with both feet on solid ground, our precious Earth home, where you always belong.

Perspective: Stop striving

Josie lives in Long Beach, California. I met her at a writer's retreat at the Grand Canyon in 2022. She spent the rest of that summer traveling and camping with her husband. She had been a restoration ecologist and is now studying to become an art therapist.

She worked in restoration ecology for years. It was her passion and calling and it was hard work, both physically and emotionally. She values traditional ecological knowledge from Indigenous knowledge holders. She is heartened by the movement she sees in professional organizations to begin to recognize this ancient wisdom.

Josie explained to me the difference between an environmentalist and an ecologist. An environmentalist has to go for the win and wants to save as much as possible. Satisfaction comes from saves and transformations. Big promises must be made to get funding for workers. An ecologist wants to crawl into the weeds and observe an ecosystem, to figure out how it works. With this understanding, she has a better grasp of how to support its thriving. Ecologists are deep listeners and deep lovers of the nonhuman world. The day-to-day work of ecology inspires wonder and awe, and feeds the soul.

Unfortunately, she couldn't tell project funders, "We took your money, and we learned a lot." She spun a good yarn, but it wasn't quite enough.

During the pandemic, she organized a team of college graduate biologists, paid near minimum wage, to do too much work with too few resources. Working with these underpaid young people, she witnessed them frantically striving to be good enough. "Did I put the plant in just the right spot? Or did I blow it?" and frantically striving to achieve a promotion to something closer to a living wage. They had internalized the rat race and were their own harshest critics. She sees the same frantic striving all over our culture: "Let's fight for a bigger piece of a smaller pie." And she saw it in herself. She was striving to meet unreasonable demands with insufficient resources, striving for justice for her employees, and striving for systemic change in her organization. It didn't happen, and finally, she refused to pay the emotional and spiritual toll. So she stopped before frantic striving destroyed her physical health. She took a year off, writing, traveling, and doing art to heal her spirit. She is beginning again in a different field, hopefully with less striving.

Preserve lessons

Every Wednesday morning, I meet a few friends at a small ecological preserve four blocks from my house. Supposedly I am working, stewarding this land. Really, I am playing, learning to know and love this little oasis in a very urban landscape.

I am the only member of our little crew under seventy. We call ourselves the Preserve Preservers. We only remove (by hand) the non-native plants that smother the natives, like black mustard, tumbleweed, and poison hemlock. I have no illusions about our accomplishments. As soon as we make headway on one species, another one rolls in and gets firmly established right under our feet. I wonder, too, how long this fragment of somewhat original landscape can persist in the face of climbing temperatures and frequent droughts. But I am not responsible for the outcome. I relax and enjoy the process.

I email the group weekly, sharing the location we'll meet and the tools we'll need. Chuck is our mentor. A retired kindergarten teacher with boundless enthusiasm, he sharpens the tools and can name most of the plants, birds, and bugs we come across. He patiently educates us on plant identification, complete with botanical names of all the parts. Repeatedly; those terms don't stick in our brains. More importantly, he coaches us on slowing down and loving the place. He sets a pace that invites renewal of body and soul. This pace also allows us to attend to the nonhuman world that surrounds us. "Look for the disconformity," Chuck likes to say. We tiptoe off-trail (with permission) to patches of artichoke thistle and tree tobacco. Along the way, we discover rare dwarf succulents, roadrunners, snakes, artful spiders, perfectly round pencil-diameter holes of ground-dwelling bees, rare cactus wrens who sound like car engines trying to start, and hawks dancing in midair.

Not all is sweetness and light. The heat wouldn't impress my inland friends, but it can be a challenge for this coastal desk jockey. I am also chigger bait; one week before I began taking precautions, I racked up 50 bites. Several of us are nursing joints that don't enjoy repeated bending and squatting. Workers charged with official maintenance of the preserve or its borders occasionally come through and mow sections, removing the natives along with the invasives. Sometimes they use herbicides, leaving bare soil ideal for a complete invasive takeover the following spring.

For me, it's all about relationships. We are forming relationships with roadrunners and hawks and pygmy blue butterflies and king-birds. We are also forming relationships with low-wage gardeners who might have little reason to love this place. Or maybe they feel the land calling too. Maybe the butterflies and the swallows flitting and diving speak to them. Whatever the reason, some of them have learned narrow leaf milkweed from tumbleweed, to spare the former and remove the latter from buffer zones adjacent to the preserve.

Newcomers to our group sometimes exhibit dogged attempts at speed and efficiency. Chuck gently ignores this behavior. He stops every ten feet along the trail to point out some insect or plant part, or to philosophize about life. I pull the newcomer aside and explain: we are learning from Chuck how to pace ourselves for the long haul so that we can know and love this place. That is more important than "getting stuff done." After all, Chuck has been stewarding wild places most days for the last twenty years. I don't aspire to that record, but I have learned to slow down, see what is in front of me, know it better from season to season and from one year to the next, and love this little preserve deeply.

Perspective: Tom's Reflections

Tom recently moved with his wife to the outskirts of the small college town of Ithaca, New York. For many years he lived, worked, and raised their children, now adults, in the more urban Brooklyn, New York. He wrote these reflections to organize his thoughts about what he is learning to value in the face of collapse.

What do I know?

Almost nothing for sure, plus I have much to unknow (which is to unlearn)

Our "pollution" includes much more than carbon and methane emissions, including the terrible no good bad stuff that we direct within and amongst ourselves (supremacies/post-enlightenment "progress"/empire)

"Us" means Beyond our species; the Thriving Life paradigm.[70]

What do I want?

To wake with sunrise, open the door, smell and hear *"the nature"*
To sip coffee with a view
To be unencumbered, spacious, exuberant, and fully rooted in the cosmos

No multitasking, no rat race, few mandatory tasks and to-do's beyond the basic needs
No traffic, no car, no rush to get to a job, no rush to be on time
No weight from things or wanting things, just the very little I need
The opportunity to be in delicious gratitude for the ever-present here-now

To walk beneath trees near a stream or beside the ocean
A view of the horizon
A bike ride, a swim, leisurely shooting basketball
Quiet walkable sidewalks to a library, a pizzeria, a grocery store
Simple meals and simple cleaning; a daily ritual

Loitering in motion
Warmth of the sun... cool evenings... warm days (no winter perhaps)

Thinking about others, caring, listening, doing, belonging
Loitering in community
Less talking and deeper conversation
Time to be sad and to be happy

And... I want the same for all.

When you slow down and attend to relationships with humans and nonhumans, what do you want?

Summary and reflection

- When you aren't sure where to go, going faster doesn't help. In this ongoing crisis, urgency may be the last thing we need. Slowing down can be healing and enlightening. I invite you to work and travel *at the speed of relationship*.

- What needs doing and what you can or should do are two entirely different things. Please don't go to pieces trying to respond to too many requests for help. How are you at setting boundaries?

- What you value is what takes your attention.

- *In hard times, people desperately need values and purpose.* They make life worth living.

- From the values and purpose you claim and your particular history, location, and passions, you will make a life. A good set of values isn't a map; it's a way of traveling.

- After pouring herself into ecological restoration work, Josie stopped striving in ways that were neither healthy nor getting the results she had hoped for. Now she is choosing a different path.

- The ladders of worth that our society invites us to climb (or condemns us for falling from) are not safe places to stand. Keeping both feet on the ground in earthy mutual relationships is much more secure and comfortable.

- From stewarding a little preserve with friends, I have learned to slow down and savor that unique part of the nonhuman world. This allows a deep connection with my fellow gardeners and with the nonhuman world.

- When you slow down and attend to relationships, what do you want? Tom made a list.

CHAPTER EIGHT

Connection and compassion

Connection and compassion are not just abstract notions. They are crucial practices during hard times. In this chapter, you will find examples and strategies for applying them when you need them most.

Compassionate witness. The power of listening deeply and holding space for people.

Perspective: Even when I'm 75. Ellen wants to offer the gift of listening well.

Scheduling connection. Don't leave it to chance.

Committing to compassion in advance. Being ready to hold our values when provoked.

Perspective: Those with the least. A story of compassion in hard times.

Jane and Sky's "living values." These are practical commitments.

The illusion of purity. And why seeking purity is a bad idea.

Perspective: Calling in. Saber is learning to do this rather than calling out.

Accepting hypocrisy. A few of my struggles with living my values.

Compassionate witness

We are all worthy of telling our stories and having them heard. We all need to be seen and honored in the same way that we all need to breathe.

– Brené Brown

Compassionate witnessing is a key practice for being the eye of the storm. I love being a compassionate witness for others. I can't count the number of times people have apologized for telling me a burden that was on their hearts or showing me their tears, as if this imposed upon me. Quite the opposite. I feel enriched and privileged to be a compassionate witness.

On Wednesday nights, a group of activists and community members in my county, many of whom have experienced imprisonment and the threat of deportation, meet for mutual aid in a group called Participatory Defense.[71] They offer support to their neighbors facing deportation or behind bars at the county jail or immigrant detention centers, and to the families of these people. The group cobbles together referrals to free law clinics, accompaniment to court hearings, and a little money for emergencies. They find leads on housing and jobs, and offer strategies to navigate the criminal legal system. They often cannot provide the help they wish they could. But they always succeed at doing one thing. They listen. Often at the end of a report from a person whose loved one is in detention, the speaker is in tears. "It's so hard. I don't know how we'll get by. But thank you. Thank you for listening. Thank you for caring."

As a transitional pastor, I would step into an existing church and try to become integral to its existing leadership and spiritual care for a year or two. One sad thing about working as the "temp" was that people did not always know or trust me well enough to allow me to listen and accompany them in times of trouble. But even an offer to listen has power. At one of my churches, a man had experienced deep setbacks during my time there. He dropped some important commitments and was seldom seen at church. I had left a couple of messages letting him know that I was available to support him, but he never responded. He did show up at my goodbye party and thanked me profusely for my support. I was confused. What support? All I did was leave voice messages on his phone. He told me, "Those messages meant so much to me. They were a lifeline. I knew somebody cared."

Being a compassionate witness means listening and caring when a person is sharing their story, sharing their heart. A compassionate witness is willing to sit with pain and trouble without trying to minimize it or brush it aside and move to action. For me, compassionate witnessing is a sacred act. I set aside my agenda and make space for someone to be heard, valued, and affirmed in their humanity and dignity. It is sometimes painful but always beautiful.

The person sharing their experience takes a risk. They want to trust that the listener will respect their truth; not judging or dismissing, not correcting, and not shutting down their vulnerable speech by immediately offering fixes or advice. Good listening is far harder to find than good advice.

I have participated in a couple of 12-step fellowships, running on the same principles as Alcoholics Anonymous. I am always amazed that a bunch of messed-up people can sit together in a room, honestly share their experiences in a simply structured meeting, and find belonging and understanding. So much relief and healing are accomplished by that simple process of listening and being heard.

In the Deep Adaptation Forum, compassionate witnessing is encouraged. One person shares what is on their heart with the group, sometimes guided by a discussion prompt. The others are reminded that listening is an active process of receiving the other person with full attention and without interruption. The listener is fully participating as well as the speaker. How is it that we witness each other's pain and come away feeling better? A burden carried alone can crush. A burden heard and acknowledged can become bearable.

When I take the role of compassionate witness, I have to believe it is not my job to fix the person I am witnessing, nor to take on their pain. My presence and acceptance of their situation is my gift to them: a powerful acknowledgment of their humanity and value. Occasionally I do take someone else's pain home. I will need to wrestle with it for a while, figuring out if I have a responsibility to act in some way. The answer is usually, "No." Then I need to grieve and let it go, because letting it crush my spirit accomplishes nothing.

Compassionate witnessing one-on-one is easy to try. Just keep listening attentively; your body language is enough. You can occasionally reflect back a few words about what you hear in compassionate terms.

Reliable compassionate listening in a group requires a *container* of structured agreements about how you will talk. Here are some groups that use compassionate listening:

- 12-step groups often take turns giving short, timed shares of "experience, strength and hope" with no "feedback, advice or crosstalk" allowed. This strict format teaches people to slow down and listen without commenting. It invites reflection and self-awareness, and it heals.

- The Way of Council from the Ojai Foundation,[72] and The Circle Way systematized by Christina Baldwin and Ann Linnea are good formats for compassionate listening. These circles "pass the talking stick" with some simple guidelines.

- Those circles are modern adaptations of the Council Circles shared by many traditional cultures. They last a very long time by the standards of my culture, often hours or days. They have stood the test of time.

- In the Deep Adaptation Forum, *Deep Listening* is a core practice.[73] Forum gatherings almost always start with a check-in for business meetings as well as support and learning meetings. "How are you, really?" We are allowed to bring our whole selves to the space.

- Conversation Cafés[74] are another format, simple and easily accessible online from the Liberating Structures menu.[75] This menu also contains other valuable group formats.

I hope you have access to groups like this in your life. Who is your compassionate witness? For whom can you be a compassionate witness?

Perspective: Even when I'm 75

Ellen lives in New Jersey, U.S. She wrote this reflection for the Deep Adaptation private Facebook group.

> I was galvanized by watching Joanna Macy, 93, and still speaking, listening, and openly weeping for the Earth. I thought, how can I, at 75, say, "I've done enough; time for me to rest?"
>
> The way she did not filter her emotion but let it hang right out there moved me in some way I cannot explain. Watching her, I absolutely "got" our predicament - in my bones, in my sinews, in my organs - in a way that I never have, even while all these years that I've been talking about collapse, I thought I had "got" it.
>
> I decided that responding to her invitation is my calling for what-ever years I have left. That calls for a big "step up" for me because by nature I am not an organizer, speaker, or initiator. Personal responsibility in this situation means that my discomfort at "putting myself out there" is not of great consequence.
>
> What am I putting myself out there to do? My decision has been informed by Joanna Macy and also by the Women's Mutual Care Circles of the Deep Adaptation Forum. I personally find the expe-rience of Deep Listening transformative. My observation of other people and hearing what they say leads me to conclude that others also find the practice deeply transformative. My experi-ence and my opinion are that we heal ourselves by listening and being listened to.
>
> I can't do anything about collapse, but I can reach out to other humans who are traumatized. Potentially I can enlarge the circles of humans who try to connect to each other. When we say, as

many do, that what remains is to love each other as best we can, that seems to be true for me.

We only have to look at the world to see that collapse is not imminent. It is happening. When a mainstream paper asks: "Has climate change made southeast Asia uninhabitable?" there's a clue. Collapse is not going to be "kumbaya"; some people will try to hold on to what they have, and they will resort to force and violence.

So, any small way that I can tilt the world toward connection, to awareness of our common humanity—that is my contribution.

Scheduling connection

Everyone deserves a good listening to. You can pay a therapist to be your good listener. But compassionate listening does not require an advanced degree, and we need not pay for it. When and with whom do you schedule connection and compassion?

I don't leave connections to chance. I schedule connecting calls using phone or Zoom (and a few visits) with about seven friends on a weekly to monthly basis. They include a neighbor, a work partner, former colleagues, a spiritual director, and friends in other countries. When I talk to them, I feel understood. I cherish these connections deeply.

My empathy buddy and I talk weekly by phone, usually for an hour. We found each other at a Nonviolent Communication retreat in 2007, and we're still at it. We check in: what's on your mind and heart right now? Then we take turns sharing a situation about which we have unresolved feelings. We receive empathy. We follow the Nonviolent Communication empathy formula, sometimes. (Are you feeling…? Are you needing…?) We are present to each other as deep listeners, always.

My empathy buddy can tell, from across the country and over the phone, when I'm not listening with my full attention. She asks me, "Are you distracted?" In this way, she has gently trained me to become aware of my own attention. We have accompanied each other in our work, with family and friends, roommates, personal struggles, you name it. With her gift of empathetic listening, I find myself clearer about a situation, more at peace, and more self-connected. At the start of a call, I often believe I have nothing going on in my life that would benefit from empathy. A little reflection reveals a little something. Or a big something.

In conversations with other friends, compassion is less structured yet reliable. If I want more focused compassionate listening than I'm getting, I might say, "I don't need to problem-solve right now. I just need to vent for a few minutes." I have learned by experience that I am better equipped to live my values when I have these conversations regularly instead of pretending that I don't need to process uncomfortable situations.

I get busy and distracted. If I waited for the phone to ring, I would seldom talk to these friends. With many of them, I am the scheduler and reminder. It's worth the trouble. So I schedule and sometimes reschedule recurring meetings, and these precious connections are not left to chance.

Committing to compassion in advance

Compassion is not a virtue—it is a commitment. It's not something we have or don't have—it's something we choose to practice.

– Brené Brown

Whatever collapse means in the future, I don't want to find excuses to make enemies of people or groups. *I commit to choosing compassion and inviting and empowering others to do the same.* Compassion is widely believed to be a weak-tea kind of practice. Showing kindness to those who appear to deserve it, and when you enjoy doing so? Easy! But reaching out toward those who terrify, repel, or infuriate us at first impulse is a deeply challenging and transformative practice.

The most memorable collapse I have witnessed was that of the twin towers of the World Trade Center in New York in 2001. It was fast and localized, though televised globally and repeatedly. And so symbolic: the demise of U.S. invulnerability. It was a tiny slap in the face of industrial consumer society but somehow a symbolic collapse of security for the United States. How did the United States respond? My government started two wars, the effects of which are still playing out over twenty years later.

Scared or unhappy people often behave badly. A stimulus that might not have bothered them in easier times can provoke them to lash out verbally or physically. They would call their behavior defense. Or, feeling overwhelmed, they may withdraw and isolate.

They may be desperate that someone *do something* to fix their unease. They may latch onto quick fixes, no matter how absurd or destructive, in an attempt to gain some power, safety, or certainty. Thus the wars in response to the twin towers collapse, and the rise of political extremism in our time.

A person's defensive behavior stimulates the people around them to respond in unhelpful ways, because it looks like an attack. So starts a feedback loop of misery creating more misery. If we keep score of wrongs, the feedback loop spins faster. Each side is furious or resentful, each thinking the other is dead wrong.

There is an easy way out. That way is scapegoating. Find a third party nobody cares about, who must be guilty of *something* if only their outsider identity, and hold *them* responsible. Then we can all feel a kind of cozy togetherness. We can share our disdain for the scapegoat, and our common defense from the threat of that scapegoat, real or imagined. For the scapegoat, our defense is an attack.

The difficult path is compassion for those who behave badly. "But they don't deserve it." That is not the point. I want to treat people well because I want to live my values, whether others deserve it or not. I frequently fail at this commitment, but I try, and that makes a difference.

You don't have to engage everyone. You can set respectful limits. Discern where your limits are. When you are willing, show compassion. Please do not attempt this until you mean it. First, you may have to step away and give *yourself* plenty of compassion for whatever injury or emotional upheaval you experienced from their behavior. Don't expect always to be able to practice compassion in the moment for those who threaten you or who trigger a reaction in you. You can commit at least to monitoring your attempts at defense and seeing how they might look like attacks to the person from whom you are defending yourself.

Here is an assortment of practices for compassion.

1. Begin by getting in touch with a source of compassion that is real and reliable to you.

Compassion is not created out of thin air. It is received from humans and from the nonhuman world, and then it can be passed on, because the source of compassion never runs out. I picture it as the sacred Source of life and love, and I envision it enveloping me in loving and healing energy. You may have a different image. I also remember the people who have inspired me with their commitment to making love real through action.

2. Think or talk through the situation.

You can write in a journal or speak with a good listener to explore what happened, how it affected you, and how you might respond. Be honest about your anger or other "negative" emotions, and your judgments of what the person or group is doing wrong. But don't stop there. Take time to realize the needs and values you cherish that are not being met in the situation.[76] How do you wish things could be different? How can you adapt to what you cannot change? What is in your control right now to further your values and meet your needs? And what is beyond your control, so that you can only grieve its lack? For instance, the past cannot be changed.

3. Try guessing stories that account for that person's behavior.

Start by assuming it's not about you. Find the basic human needs they were trying to meet when they did what didn't work for you. Even if what they did was outrageous, at the root of their behavior is a universal human need they wanted to meet.

4. Pray or meditate for that person's well-being.

Or, if that is a leap too far, you could pray for your own peace of mind. One form of prayer for someone I resent is asking for that person to receive the good things I want for myself. Twelve-steppers are instructed to pray in this way for people against whom they carry resentment, daily for two weeks, repeating as necessary. Another option is the Buddhist loving-kindness meditation. This meditation can look something like this:

May you be safe.
May you be happy.
May you be healthy.
May you be at peace.

The more absurd this approach seems to me, the more I need it.

In situations of chaos and fear, seemingly reasonable responses of defense or punishment can take a community into a spiral of disconnection and even violence. To interrupt this spiral, we need to be prepared. We have to commit in advance to act from compassion when our instincts tell us otherwise. Our defenses and reactions almost always make things worse. We can start practicing compassion with people on the street, neighbors, store clerks, and anyone we meet. And we can take courage from the models of oppressed people who have chosen nonviolent direct action instead of violence, including Gandhi, the Standing Rock Water Protectors, Rigoberta Menchú, and hundreds more.[77]

Perspective: Those with the least

Author Margi Prideaux retells a story of practical compassion shared by a friend who survived the Australian Lismore floods of early 2022. In Margi's words:

Water reached the rooftops of two-story buildings. It was deep, fast-moving, and extremely dangerous. The government was completely negligent. They didn't turn up for four days. Most of the rescues were done by the community, in little boats going around just listening for people crying out for help from inside their roofs. When the government did turn up, they made people go through crazy levels of paperwork and didn't come with any food or water.

A large area was flooded, and in one township was a First Nations newspaper. These were people who had been to hell and back for generations. They have every reason to just turn their backs on everyone with white skin. Their newspaper building was above the flood point. They ceased the paper and pushed all of the equipment out of the way. They got their community network to bring in water and food and clothes. And they opened a drop-in center within about two days. They went around in boats and dropped off food and water to people who had stayed in their properties. After the first week, they brought in counselors, doctors, nurses, trauma specialists for children, and more. They used the newspaper's own not-for-profit funds. They set up rooms where you could have a massage, you could have acupuncture for stress relief, you could see an herbalist… they made the entire building into a rescue and relief center that welcomed everyone.

They knew. Their world had been prolonged trauma, so they just immediately knew what people would need. And they kept providing relief for about four months, with no government funding. They don't want attention; they don't want to do interviews. Amazing.

Those with the least sometimes do the most.

Perspective: Living our values

In 2012 Jane and Sky wrote this simple list of practical values they planned to live out in their Vermont home with its large garden: settling in one place, making their home a refuge for others, and sharing food. Things didn't work out as planned, but they were still guided by these values.

Living Values:

Going deeper with stuff that's important

Food
Prep for the future
Long term projects

Settling (living in one place)
Know where we are
Others know where we are

Use stability for
The greater good
Our good
Our friends' and neighbors' good
The world's good

Doing stuff with other people
Using our house and land for that

Local is global

Earth is in rough shape

Creative things we do here help out there

What are your Living Values? Or what might you like them to be?

The illusion of purity

The world is in a terrible mess. It is toxic, irradiated, and full of injustice. Aiming to stand aside from the mess can produce a seemingly satisfying self-righteousness in the scant moments we achieve it, but since it is ultimately impossible, individual purity will always disappoint. Might it be better to understand complexity and, indeed, our own complicity in much of what we think of as bad, as fundamental to our lives? … the only answer—if we are to have any hope of tackling the past, present, and future of colonialism, disease, pollution, and climate change—is a resounding yes.

– Alexis Shotwell[78]

I invite you to free yourself from the illusion that the right actions can make anyone pure. And see if you can catch yourself before you shun or shame people who are not pure enough. If someone demands that everyone must follow their path, suspect a purity mindset at work.

Even if purity were possible in our complex predicament, purity means putting up a wall with the pure on one side and the rest of us on the other. A person seeking purity may not intend to shame, shun or exclude those who live differently, but it happens regardless of their intentions. If we share a heartfelt desire to live differently, people might be charmed by our eccentric ways of loving people or the planet. If we demand they behave like us, they probably will get turned off by their sense of being judged. Our insistence on purity might occasionally convince someone to join us. More often, we will leave people annoyed and unwilling to listen to us.

Someone asked me whether we could mandate the actions needed to save the planet (whatever that means) by having an authoritarian government. Seriously. That is where purity can take us.

I am not trying to discourage anyone from taking action. The same action can be done for any number of reasons. I want to examine my motives and make sure they are not about being pure, right, or above reproach. A good test is: can I befriend somebody who doesn't do what I'm doing? Someone who doesn't even understand what I'm doing and doesn't want to hear about it?

Here are some choices that can get mixed up in purity and "one right way." As I explain them, I feel my judgment landing on those who would be pure. Nobody's perfect.

Eating vegan

Industrial animal agriculture is almost always land- and energy-intensive, horrifically polluting, and heartbreakingly inhumane. Advocating to regulate it and refusing to participate in it make sense. Purity goes further, demands we all become vegan, and shames us for eating animals.

Animals killing animals for food is part of the cycle of life, and humans have been hunters or kept food animals in almost every traditional culture. The moral relationship of humans to animals, respecting the gifts of their lives to feed ours, is addressed in many traditional cultures with rituals and expressions of gratitude. There is not one right way.

Eating organic

Do you have the money and cooking facilities for such a choice? Lucky you! Are you aware that "organic" is a label with many different regulatory meanings in various legal jurisdictions? Some are more helpful than others. Many small farmers can't afford the certification. Some "organic" pesticides can put people and animals at greater risk than the synthetic substances they replace.

Another option that is more work but may have more substance is to shorten the supply chain. This might mean eating locally and developing relationships with growers at farmer's markets or through community-supported agriculture, CSAs. This usually means signing up for a box of food each week or two from a local farmer. We can learn how they farm, labels aside, and support them with our food budget. Do have empathy for people who choose the cheapest or fastest food. It is not always much of a choice. There is not one right way.

Not having children or having fewer children

Wealthy and well-educated people know they have choices about whether to have children or how many to have. Many people don't have those choices or don't know how to access them. Historically, people have assigned great meaning to children and the legacy they represent. In a society without a social safety net, children are your pension. Having several children is only prudent in a society where child death rates are (or have recently been) high. So this kind of purity can easily lead, intentionally or not, to racism and classism. To people with children, purity about not having children sounds like the devaluing of children. When given access to birth control, most women will have only the number of children they can care for well, not more. Religious restrictions on reproductive health have issues of purity too, and those restrictions infuriate me. There is not one right way.

Living without a car, without modern gadgets, without flying, off the grid…

Opting out of parts of industrial consumer society can be pro-foundly life-affirming. Many people don't have the awareness, means, or self-will to do that. Can we find ways to change social pressures and government policies to encourage living more simply instead of shunning or shaming people who are engulfed by an addictive system? Meanwhile, we can start a conversation about our offbeat choices. The more I respect a person, the more I find myself emulating them. There is not one right way.

Coercive advocacy

Sign up for my political action, support my tech fix, buy my miracle product, or shame on you, blame yourself for the state of the planet. Take these steps to save the planet, or else!

Desperate people can get angry and coercive. A friend who was deeply involved in student Civil Rights movements in the 1960s told me that the angriest people he ever met were peace activists. It's easy to become like what we hate by fighting it. There is not one right way.

Not participating in organizations, or with people, who don't see things exactly the way you do.

This may look like calling out people who are not sufficiently "woke," or one of a hundred other criteria. It may look like boycotting elections because neither party is pure enough, avoiding work that is not totally aligned with your values, or shunning people who don't share your values.

I am formed by unjust systems and embedded in unjust systems; there is nowhere for me to stand free from participating in oppression. And informed people do disagree about which personal and political choices are the wisest.

Loretta J. Ross has advised that if you share about 50% of someone's agenda and values, you are wise to set the rest aside and work with them on the common ground.[79] There is not one right way.

—◊◊◊—

Modern people are entangled in life-denying systems. Purity may be an attempt to avoid the crushing guilt that can come with participating in the systems that have created our predicament. The alternative to purity is to learn to live with ambiguity and show compassion for ourselves and all the others entangled in those systems. This includes almost everyone I know. We are all human, broken, complicit, and also capable of great compassion and generosity.

Nobody is more dangerous than he who imagines himself pure in heart; for his purity, by definition, is unassailable.

— James Baldwin

Perspective: Calling in

Saber is a student at the University of Pretoria in South Africa and uses they/them pronouns. They value justice for women and people of color in a country that is still profoundly unjust. They want to invite people to become aware of the effects of their words in the gentle and respectful way they learned from a dear school friend Keabetswe, nicknamed Kea.

Saber knows that when they get defensive about something someone has said, the speaker immediately backs off, and the opportunity to make a connection is lost. So when they want to have powerful discussions, they know how important it is to keep their emotions settled. In high school, Kea would challenge Saber's best friend because he didn't respect some people the way Kea wanted. But Kea would always do it in a very kind and helpful manner. She practiced *calling in* instead of calling out.[80] She would say, "Hey, that didn't come across quite right; maybe say it like this."

Since they left school, the friend who was called in has spoken at length about how much he appreciated Kea's gentle invitations to respect. Kea was willing to sit down and have long conversations with him and help him get to a point where he could think differently. The friend now practices Kea's approach to broaching hard topics. These conversations have inspired Saber to try Kea's approach. Instead of just saying, "No, you can't say that, that's so sexist," or, "Don't do that. You're going to kill the environment, don't eat meat; it's so bad," Saber will invite people to sit down with them. If the person is willing to have the conversation, it will be calm, friendly, and usually happy and helpful on both sides. Maybe they won't agree on specific points. But Saber doesn't need to say, "I think you're wrong." They can say, "This is how I go about it, and this is why." Sometimes they change a person's mind, or maybe the person returns to them in a few weeks and says, "I was thinking about that. And actually, I think I agree with you." At the least, it's given that person something to think about.

Saber has used this approach when discussing controversial topics with their mom. Their mom in turn reported that she now uses the same approach at work with her colleagues. Saber notes that interacting in this way seems to have a domino effect, and that adds to its beauty. Those moments of interaction leave them feeling uplifted and hopeful about the future. When the words are kind, calm, and respectful, people are willing to listen, and their minds may be changed.

Calling out often feels to the recipient like shaming and rejecting. The term *calling in* comes from the work of Loretta J. Ross.[81] She uses it to describe the kind of caring conversation Saber is practicing.

Accepting hypocrisy

If no hypocrites were permitted to hold opinions, there would likely be no opinions at all.

– Shaun Chamberlin[82]

Since none of us are pure, we might as well accept our hypocrisy. I don't enjoy mine, but I'm usually able to laugh at it. Once I get over the illusion of purity, I am freed to work within my limitations.

—————

I have a fraught relationship with a "Carolina Oak" (species undetermined) planted on a rise in the common area behind my backyard. It is about 25 feet high and 40 feet wide, rather rectangular. Its stately canopy shades the only sunny spot in my backyard all winter when my lemon tree and white sage are trying to grow. This oak was chosen by the common area management staff instead of an oak species native to California because California oaks tend to die when overwatered, and overwatering is what California homeowner's associations do. This oak is the home of wacky non-native squirrels who happily plant its acorns in my yard. I get dozens of seedlings. In addition to digging up my yard, the squirrels eat the fruit off my trees and taunt my cat. I have to admit it's entertaining. When one bounteous branch of this mighty oak grew horizontally over the wall and into my yard, I got a friend to cut that branch off with a chainsaw. The association manager chewed me out. Nothing new there.

I am doing my best to make peace with this oak. It is beautiful, if a couple of thousand miles out of place. It didn't ask to be planted here; it's doing its best to be its stately self. I am not a Native either. That oak is the largest tree on the street. If it weren't behind a wall, it might be an excellent shade tree for lounging in hot weather. Perhaps someday I will put a gate in the wall and set some lawn chairs under its canopy. Perhaps someday I will be grateful for its acorns. Perhaps someday I will mourn its death when suburban irrigation ends.

—✻—

I have a fraught relationship with my car. I live in endless suburbia, and I have some bad joints that make bicycling iffy, so I hang onto it. Also, I'm lazy. Also, I'm scared of "death by automobile," as happens to bicycle riders with disturbing regularity in my community.

Most of my car's miles are spent driving to see my mother. There are no buses or trains for the 300 miles from Irvine to Santa Cruz. I could afford an electric car, but could I charge it reliably between here and there? Would I be using much less fossil fuel buying a new car? I know I am deep in the industrial consumer machine, and I am not willing to do the work, nor alienate myself from family and friends, to try to extract myself much further than the small efforts I am making.

—✻—

I have a fraught relationship with the environmentalists in my neighborhood. They are many, and they mean well. When my city put out a call to create block-by-block "Cool City" climate action groups, most of the participants turned out to be my neighbors. They are crusading to replace natural gas home appliances with electric, install solar panels, and more. They might be my people if the organizers didn't need to pretend that, "It's not too late to fix the climate crisis." After some initial conversations, the organizers and I agreed that I would not be a fit. I was dubbed "Debbie Downer" by one trainee. This is not a title I relish, but I have trouble holding my tongue about the naïveté and futility of some of their proposed actions.

I miss being with them. I meet my kindred spirits online, but people in the flesh come with real advantages. I have chosen not to invest time with them for now, but I want to open my heart to them, whatever I think of their strategies. I questioned one neighbor about the limitations of what the groups were undertaking. Her answer touched me. "My teenage daughter is asking what we are doing to save the planet. I have to do something, for her, with her. I can't do nothing." What am I doing for the children, I wonder? Not much. What could I do? I continue to hold that question in my heart.

—◦◦◦—

I want to open my heart, living gratefully, generously, and with gratitude. When I do, connection with people and the nonhuman world matter more, and stuff and striving matter less. My time may be best spent practicing this approach rather than trying not to be a hypocrite.

Summary and reflection

- *Compassionate witnessing is a key practice for being the eye of the storm.* This means not fixing or judging, but attending, accepting, and caring.

- Who is your compassionate witness? For whom can you be a compassionate witness?

- With Joanna Macy as her inspiration, Ellen is determined to give others the same compassionate listening she has received.

- You can pay a therapist to be your good listener. But compassionate listening does not require an advanced degree, and we need not pay for it. I do need to schedule it in advance. When and with whom do you schedule connection and compassion?

- Miserable people often behave badly. Scapegoating makes things worse. Defense looks to the other like attack, and so begins the downward spiral. Responding compassionately is a commitment we choose in advance of the provocation.

- I commit to choosing compassion now for when it gets hard, and inviting and empowering others to do the same.

- I offer five practices for getting to a mindset of compassion.

- A group of Indigenous Australians had every right to be bitter toward the descendants of settlers. Instead, after a disastrous flood, they showed the practical compassion their white neighbors needed.

- Jane and Sky wrote their "living values." What are your living values? Or what might you like them to be?

- Free yourself of the illusion that your right actions will make you pure. And see if you can catch yourself before you shun or shame people who make choices different from yours.

- Some topics related to collapse that can get mixed up in purity thinking include eating vegan, eating organic, not having children, living without certain modern conveniences, and coercive advocacy.

- The alternative to purity is to learn to live with ambiguity and show compassion for ourselves and all the others entangled in life-denying systems. This includes almost everyone I know.

- Saber is mastering the art of *calling in*. When the words are kind, calm, and respectful, people are willing to listen, and minds can be changed.

- I am a hypocrite. I am not trying to be pure. I am trying to be generous and grateful, and that is a lot more fun.

CHAPTER NINE

Letting go

In this chapter, we start getting our hands dirty and having fun. Imagine letting go of industrial consumer society. What might that look like? What can you let go of temporarily to experiment with living more simply? Traveling light in the storm is a challenge, and also a gift.

The illusion of security. We are vulnerable; best to face that reality.

Perspective: Impermanence, the immersion course. Carla is a Buddhist teacher. Her house burnt down.

Stone soup. A vision for hard times.

A ritual for letting go. With fire. Everybody loves it!

Letting go as solidarity and eco-cultural healing. The upside of living simply.

Perspective: Coming home. Jay has let go of a career-focused lifestyle.

Plan B. For when Plan A doesn't happen.

Letting go for living together. Living together means letting go of getting your way.

Perspective: Letting go many times. In Kat's career, she has let go of many jobs. That is something you might have to do if you want to be paid to follow your passion.

Letting go as freedom. Living simply as freedom from fear, from bills, and from encumbrance.

Perspective: Maple syrup for oranges and sunshine. Jane is an old hand at living simply.

Many voices: Simple living. People do it in all kinds of ways.

The illusion of security

Become a tree.
Your limbs no longer move under their own power.
Your lower limbs, your roots, are quite stuck
but fractal; they move by entangling, they explore.
Your upper limbs dance but not at your command
in the wind.
They bend, or they break.
You face the storm, and endure.
And when you no longer endure, you stand.
Weathered cenotaph,
home to birds of prey.

– Terry

We who benefit from modernity can structure and plan our lives with the comforting illusion of security. When that curtain of illusion is pulled back, the vulnerability of what we hold dear is revealed. If this is your situation, I suggest that you get used to peeking behind the curtain. This is not to burden you with anxiety from visions of future catastrophe, but to gently begin the process of living with loss by imagining that loss. You can also be in solidarity with those for whom loss is already a reality. You know your own heart and your nervous system. Make your own choices about this.

Sometimes the illusion of security holds well. Invariably, this illusion distances us from those who are experiencing loss. The illusion of security has held for many people in my country for over half a century. The treasured pension plans and paid-off mortgages of the second half of the twentieth century are providing security to some retired people now. The possibility of the stock market crashing or their pension failing hides behind a curtain they dare not draw back. Meanwhile, some people are noticing the rise of authoritarianism and the loss of human rights in countries where political security was taken for granted, including mine. Some are witnessing economic losses and inequities, housing shortages, the erosion of affordable medical services, wars and the threat of nuclear war, refugee crises, and crop failures. Others are living through these things now.

I feel reassured when I remember that most people in times and places other than mine have had a fraction of my material comforts and security. Many of them had first-hand experience of living through devastating losses. Those who survived learned that security is never assured. They found ways to navigate their losses, physically, emotionally, and spiritually.

Traditional cultures and religions have practiced various strategies to face our vulnerability. Fasting, physical trials, vigils alone in wilderness or monastery, and meditation on one's death or existence before or after one's lifetime: these are some ways that people have traditionally confronted the illusion of security. The skull on the scholar's desk in many Renaissance paintings is a *memento mori*, a reminder of one's mortality. Stories and tools from traditional spirituality or religions helped people to face uncertainty, fear, and loss. It is time to find stories and tools for our age of loss, new or old. Cult leaders and autocrats will be offering theirs. More just and respectful resources to deal with vulnerability will take some exploration and patience, but they are worth crafting and sharing.

Perspective: Impermanence, the immersion course

In 2020, Carla's home of 18 years and all her belongings were destroyed in a massive complex of fires that ravaged the hills just north of Santa Cruz, California, about 40 miles south of San Francisco. Carla is a longtime teacher of Buddhism, and one of the central tenets of Buddhism is impermanence. She appreciates the irony of her struggles with loss, identity, and belief, in the aftermath of this disaster.

Carla describes the early days after the fire as a prolonged period of shock. Even the wisest teachings are subject to bodily limits. She devised a plan to hop in a recreational vehicle and visit around the country, deciding where to resettle. Health challenges got in the way. She took great comfort from caring people who offered their homes for a time and from people who truly understood what it was like to lose all one's possessions. She also took comfort in the things that Buddhist teachings say persist: wisdom, compassion, love, and connection.

She points out that there is a right time and place for certain teachings. The Buddhist teaching that difficulties are opportunities for learning and deepening? This teaching is probably the worst thing to say to someone who has just experienced a loss like hers. Her spiritual teachings do not mean she gets to bypass anxiety, pain, or grief. Grief has become a core process for her, one she is still learning.

Carla's beliefs and plans remained unsettled two years after the fire when I interviewed her. At publication, she is still living with her partner in a travel trailer, seeking a new place for the two of them to settle. In this nomadic life, her familiar daily meditation practice reliably brings her a kind of direct wisdom. It also brings her "home" in some deeply felt sense, no small thing when her physical home is gone. Taking in what is before her in her meditation, letting go of thoughts about all that was lost or remains unsettled or uncertain for the future, reliably brings her moments that are "almost always pretty fine."

Stone soup

In a vision, I am cooking stone soup for a large group over an open fire. I am crying. I desperately want to feed them something better than thin broth, something with enough fuel to keep them going. Then we seat ourselves in a circle around the fire, waiting for our stone soup…

A golden man with golden braids appears to us and says:

I wish I had barley for your soup but I don't. I have a story.

The story is about the people who let go.

There is so much beauty in the world and so much terror in the world,

and so much insistence that things not change and that we keep hurtling ourselves off a cliff.

But you folks are letting go.

I'm so proud of you.

You are practicing letting go and it's not easy work

but it's necessary work and everybody will be doing it soon.

And you will know how to let go.

You are letting go of comforts.

And you are letting go of old stories that don't fit anymore.

And you are letting go of lies.

And you are letting go of insanity.

And you don't know what's next, and that's OK.

I will stay with you until you know what's next.

You will get through this time.

And your stomachs will be full again.

And the golden man invites us to dance.

And when he dances with you, he grabs your shoulders, and he takes away your pain, at least for a little while,

Thank you, golden man with the golden braids.

And it might seem ridiculous to dance when you're hungry but it's actually a pretty good idea.

So it was that our stone soup was seasoned well.

A ritual for letting go

I introduced this ritual to my churches for Ash Wednesday, the first day of the pre-Easter season of Lent. People always appreciated it, and it became an annual observance in some churches. You can do it alone, but a shared ritual is more powerful.

To prepare,
- **Slips of paper** (about 4x5 inches, or 1/4 of a regular A4 or notebook sheet)
- **Pencils or pens** and something flat to write on
- *If indoors:* a **fireplace**, or good ventilation and a high ceiling, and a large shallow **fireproof bowl** (~16-inch) metal or glass bowl
- *If outdoors:* no wind, **fire pit**
- **Long lighter** (make sure in advance that the person knows how to work it reliably)
- Optional: **soft charcoal**, crushed, in a **small dish**, and a **spoon**
- Be sure to have a **fire extinguisher** ready if you are indoors

To do the ritual:
- Each person takes a slip of paper, a writing instrument, and something flat to write on.

- Invite people to reflect on what they are letting go or want to let go. Poetry and music are often more effective than long speeches for ritual preparation.

- They write that thing or things on the paper slip.

- They crumple the paper gently into a loose ball. (Demonstrate; flat or crushed-too-tight papers won't burn well.)

- One by one, people come forward to put their paper balls in the large bowl.

- With the long lighter, ignite the papers in the bowl. Allow them to burn.

For Ash Wednesday, the ritual continues; feel free to skip or modify:

- Mix a spoonful of the paper ash with a bit of crushed soft charcoal in the small dish.

- The host marks the forehead of each person who wishes it, saying the traditional words: "You are dust, and to dust you shall return," or other words like, "You are Earth, and to Earth you shall return," "You are dust and Spirit, a beloved child of God," or whatever is appropriate to the occasion.

- Closing words: for instance, an affirmation or blessing, or a suitable poem.

Rituals and embodied practices go beyond thoughts and words to lived experience, allowing us to be touched at deeper levels than words can reach. A rich trove of rituals honoring Earth and making room for all feelings resides on the Work That Reconnects website.[83]

Letting go as solidarity and eco-cultural healing

Reporter to Gandhi: "What do you think of Western civilization?"
Gandhi to reporter: "I think it would be a good idea."[84]

Many of the losses we may face on a personal level in a societal decline (or collapse) are of things people outside industrial consumer societies never expected to have. Some of these are comforts. Some are entertainment. Some prolong life and health, if you have the money. Surely, we can be in solidarity with most of the human species if it becomes our turn to do without these things. What can you let go of temporarily to experiment with living more simply? Don't make it a sacrifice. Let it be your solidarity and freedom.

In a spirit of play and exploration, we who have privilege can try doing without some things we don't need, at least for a short time. I am not inviting you to try this to become pure from the taint of modernity or to earn virtue in saving a portion of the nonhuman world. I want us to learn what we truly need and value. We may also join in solidarity with those who have already lost, or never had, what we can freely choose to let go.

The pandemic taught me that I don't need to drive much. During most of 2020 and 2021, I drove to the grocery store once a week, to medical appointments, and to visit my mother a few times a year. The other trips I made I could count on my fingers. I realized I didn't need to resume most of the driving I had done in the past. This was hardly a sacrifice for me. I have added back church, two short trips to a pub and a store per week, and not much else. I notice that my world is smaller in a way that I find helpful. This focus helps me appreciate my home and neighborhood, instead of wandering around for entertainment or distraction. With the internet, my world is not so small.

If you drive, what would happen if you drove less? Can you get exercise and enjoyment without traveling to a gym, court, or specialized classes? How few powered devices do you need? How few plastics? How little garbage can you produce in a week? How little money can you spend? Can you live out of a suitcase worth of clothes and sundries? Can you entertain yourself without electronics? By experimenting with one or two of these challenges, you will prepare yourself in a small way for future losses of industrial consumer comforts. As a happy by-product, you may find it easy or enjoyable to let go of some of those things now.

What would you eat if you couldn't go to the grocery store for a few weeks? What would you eat if you only ate locally-grown foods? What do you eat that you could grow yourself?

What would living in your home be like without electricity, plumbing, heating, or cooling? Camping at home? Why not? How might you make that comfortable? Can you make it a game? Where would you store and empty the commode? Yuck. But anyone caring for an invalid has survived the experience. Outdoor and camping stores and "prepping" websites have lots of pricey gadgets to sell you. Be careful that "letting go" does not become a reason to buy new equipment and gadgets, aside from some emergency basics.

I am frightened by the number of people I know who own two homes, while so many people in my area can't afford one home. How did we get so out of balance? Occasionally, can you use your home(s) to shelter others? If you were to use it to host as many people as is common outside industrialized countries, that would typically be one or two rooms per family. Too cozy? I agree. I'm noticing that what I take for granted is a luxury to much of the world. Maybe you could offer a spare bedroom under the market price to a college student who needs shelter: temporary and very simple in their expectations. You could offer or use a free homestay through various organizations on the web. And I hope it goes without saying that you could open your home to refugees.

You could try the "buy nothing challenge" for a week, a month, or (I am told most effectively) a year! You could try a zero-waste challenge for a fixed period as well. In all these experiments, you will discover what is easy to let go of and what is hard. You can learn how to deal with the pangs of loss. It's all about the stories we tell ourselves.

As you have been reading this, pay attention: what sounds doable or exciting to you? What sounds scary or just impossible? Please remember to try any experiments in letting go *with a playful spirit.* Not to prove virtue, relieve guilt, or fix anything, but to learn what it's like to live more simply and develop the muscles for letting go. Then when it becomes necessary or compelling, you will know how to let go of something. In the meantime, you will have a bit of confidence that comes from knowing what to expect.

I have just given you more than a year's worth of exploration. I haven't tried most of these letting-go experiments myself. What appeals? If you are new to simple living, start with one thing. See if you can entice a buddy or a group to join you. You can research options, learn together, and share your struggles and laughs.

Perspective: Coming home

Jay lives in eastern Australia. We met her in chapter six. Recently she let go of a busy career and a continent-hopping lifestyle. Jay and her partner had leapfrogged nationally and internationally for a dozen years, one finding a job in a new city and the other following a few months later. She had defined this as "success." She hadn't pictured herself settling down. Yet she realized that, in the face of great uncertainty about the future, being near family made sense, and having a home and a garden could make staying in place enjoyable.

Jay's partner got a job in a small town nearer to their families and they bought a house there. She and her partner are currently living on his income. She is near national parks, the ocean, and public transit. Still newish, she is immersing herself in the neighborhood and local community as she always does after a move. She has time to visit her family at leisure, not just one hectic visit a year.

She is realizing that her career is not the priority it used to be. Work until retirement? Is she going to have a retirement? Maybe she should enjoy today while her partner's job supports their living costs. This was a challenge, letting go of her assumptions that a modern woman should have a career. She is not currently seeking a paid job, instead taking time to decide who she is and what she will do in these strange times. Don't imagine she is idle; she has been signing up for all kinds of volunteer projects that happen to be fitted for hard times: emergency rescue services, wilderness care, teaching immigrants English… The list is long. She feels useful. Probably her old habits of keeping busy are playing out in new ways. She is guiding "Forest Therapy" walks as her new vocation, not a lucrative one. And that's OK.

Plan B

Letting go is easier when we remember that we usually have more options than we realize. When Plan A fails, Plan B is waiting to be discovered. The trick is to be flexible enough to look for Plan B. Otherwise, we will be tempted to cling to Plan A as the only way, and vulnerable to turning bitter, belligerent, or despairing when Plan A is lost. Even better if we can consider a Plan B or C when our Plan A has not yet failed. This is true for emergency preparedness but also for food choices, choosing where our kids go to school, making a home, and making meaning in our lives.

Letting go of our expectation that things will go according to plan means letting go of getting our way. Privileged people have not faced this challenge very often. It is a practice we do well to master.

Of course, we want Plan A. But our comfort or security does not require one and only one plan (or home, or person, or job…) Allowing yourself to imagine a Plan B may require some emotional work and some grieving. In a time of decline, it may be hard on the ego to let go of things that shaped our identity or status. We may have to do something that doesn't fit our self-image. We may have to ask for what we need to live. A certain self-image and perceived independence are two more things we might need to let go of. Resorting to our Plan B may challenge us deeply, but it will also leave us more creative and more flexible, and thus resilient.

Letting go for living together

Those who live with family or in an intentional community practice a daily give-and-take. Who decides what? Who is responsible for what? We don't get our Plan A when the other household members don't follow that plan. Thinking it should be otherwise is another luxury that industrial consumer society might have taught us to expect. Individual choice is always subject to negotiation with the choices of others. Holding my choices loosely, paradoxically, allows more room for negotiating my best outcome with my spouse. Scott doesn't like being told what to do. As we explore our feelings and ideas, we discover new options. As I experiment, he may find some of my choices aren't as bad as he thought... and I find the same.

How much do your experiments in letting go impact other members of your household? Are you willing to let go of their buy-in to your plan? Instead, can you try some lifestyle changes solo while you are still in the relationship and household? You can probably do a lot of experimenting solo before making any drastic changes that your living partner(s) resist. You may even be able to recruit their participation for a trial run that, you assure them, isn't permanent. Moving house, for instance, is a big decision that can usually be given a test run. If you are still opposing your partner, you may want to examine your motivation. What is the urgency of your Plan A? What result are you seeking? Must you be right or pure? Do you imagine you can escape industrial consumer society? Are you demanding that others think and act like you or approve of all your choices?

It is sweet when our whole household shares our values. But it is also sweet when one family member can support the other in doing things that are not their first choice out of care and consideration for the other. My partner not letting go of some aspect of modernity will not tip the planetary balance. His accommodation of my choices, and his support of choices I've made that he would never have made for himself, have made this book possible.

Perspective: Letting go many times

Kat was raised in Australia, moved to England, and now lives in Scotland. Her story began in Chapter One. Here she tells of her meandering work history, charting her course by feel, repeatedly letting go of jobs that didn't serve life, finding her people, and becoming an umbrella. This is the longest section in the book. It might serve as a guide for those working within a system for change: be prepared to change course repeatedly, and bring your umbrella! In her words:

> When I finished my education, I was wildly enthusiastic about the possibility to have a real impact. Finally, I could have some power and influence. The chairman of the [Australian] Environmental Protection Agency, whose name was Barry Carbon, which still makes me laugh…. he told the whole agency team, "I need you all to go from here and make mistakes. Because if you're not making mistakes, then you're doing things the way they've always been done. And the way things have always been done is not getting us anywhere. So it's time to start pushing those boundaries and trying new things."

> And then little by little, although I didn't see it at the time, I became disenfranchised or I lost faith. I was gathering information on and observing the natural environment. But none of that information was translating into any meaningful policy or any meaningful direction of travel. So I moved on from being a researcher.

> I meandered into policy, thinking, "Well, if it's not about the data, maybe it's about how that data is applied to policy or the development of legislation." But I realized, after a period of doing that work, that actually, that wasn't where the impact was either. So then I migrated into regulation. I thought, "Maybe it's about regulation, maybe we just need to stop the people doing the things that are wrong, and we need to find them, and that's how

we interrupt these patterns of extractivism and disrespect and disregard of the environment." So I worked around enforcement for a period and then realized that wasn't making any earthly difference at all. We were only penalizing and punishing little operators who weren't having much impact. When we prosecuted large organizations, they factor that sh*t into their balance sheet. If they're fined a quarter of a million dollars, that's neither here nor there to them. They just write it in their budget. Then I understood that enforcement wasn't going to work, either.

At this point, I went back to my first love, which was research. But in the time that I'd been away, the budgets that were available had shrunk by an entire order of magnitude. We couldn't cover the area that we were covering. The questions that needed answering weren't getting any smaller. So we got really canny about how to collect information. We started what I came to understand later was a first effort in community science. We worked with landholders to collect samples. Instead of my team driving from point to point on the river and physically collecting the samples, they were like postmen. They drove to the end of each driveway and collected the little cooler full of samples that the farmer had collected in the morning. We were able to determine patterns: in the way that water quality was changing, in the way that soil qualities were changing.

During that period, I still had the arrogance that comes with being a scientist, thank you very much. "We're here to help." Until the farmers started asking questions and landowners and the community started saying, "What is this information? And what is it telling you? And what does that mean? And why does that happen?" So we started having to work much more closely with the community. "All we can tell you is that there's a huge spike in phosphate concentrations this month from this area." They said, "Oh, that's when such and such ripped out all their grapevines," or, "That's when they logged that piece of forest at the top of the hill." What's really powerful is giving people access to that information and allowing it to inform their daily choices in terms of… When do they apply their fertilizers? When did they plow? Do they plow at all? Do they shift away from agronomy into another thing? That connection started to feel really powerful.

My discontent with being part of a government agency really set in at that point because my allegiance, I suppose my empathy, my connection, was with those people outside of the agency who were taking this powerful stuff and it was making a difference. Maybe not at a global scale, but it was making an enormous difference right there on the ground. I broke a lot of rules in terms of sharing information. My personal values and my personal ethics were such that I would only ever be very transparent and very open. Government agencies don't like that. There's a line to be toed. There's a script to be followed, and I wouldn't follow it. My tenure in government became vulnerable then. I would get personal telephone calls from chief execs of the agency telling me to wind my neck in, or why did I give away that piece of information? Why was I doing this? Why was I doing that? I wasn't actually doing anything that would warrant being fired, so they slowly clipped my wings to the point where I just sat at a desk crunching numbers. My life became a bit soul-destroying.

Next, I was given the invitation to work in a private consultancy. We specialized in an innovative, cutting-edge niche. We weren't writing environmental risk assessments; we weren't working with large multinational companies around how they get permission to rape and pillage the planet. We were working with really wicked challenges that the conventional system had no solution for. We designed and delivered the first entirely passive waste-water treatment system in the Southern Hemisphere, for a community on an island that was hard to get to. We designed concepts for water treatment and reuse at the point of origin, and treating water passively to better than drinking water standards, without the need for infrastructure, without the need for highly technical, trained officers, without the need for constant energy input, without a need for chemicals. That was really exciting stuff to be part of: extraordinary projects, making enormous differences to the communities where they were being implemented.

Then I made the decision to move back to the UK with my family, which coincided with the global financial crisis in 2008. I wanted my daughter to connect with her ancestry and her heritage, and I wanted my mum to be at peace and comfortable. I made a decision at that point that I wouldn't work full-time anymore, that there was more to life. I wouldn't sacrifice the quality of life for quantity. I would have less and be more; that was the decision. I would only ever work enough to pay my bills. The rest of my time I would give away where it could be of value to people who couldn't afford to pay for the skills of people like me to make a difference in their life. That's the pathway I've been on for a long time.

When I got back to England, I realized that I was actually quite Australian, and I felt the oppressive and repressive nature of British culture. Someone wise from Europe said to me, "British people are too polite to be honest." I was back to feeling like we can't make a difference in this system. It's impenetrable. I solved that problem by working exclusively overseas. For a few years of my life, I was flying. I was spending every third month in Australia continuing to work with the consultancy that had gifted me with so many incredible opportunities. And half of every other month I was consulting across Central Europe: Romania, Slovakia, Slovenia, Croatia, and a little bit of time in Norway. I didn't work in England at all.

As my daughter got to her teenage years, I realized that I couldn't continue to live the way I had done because she needed me to be much more present. This wasn't about her needing to be cared for or fed or clothed, or the physical day-to-day. She needed my presence for her mental and spiritual well-being. Something had to change, so I joined the community and voluntary sector. I've done government, I've done private. The community and voluntary sector is what I've been doing now for almost a decade, part-time.

When I took over the organization, it was at the point of failure. I didn't expect to be offered the job of the boss. It was a catchment scientist that they were looking for. I submitted an application. I drove two hours to the interview. I sat in front of a panel of three white men who asked me a lot of intense questions. About half-way through the interview, they stopped me. They said, "I'm really sorry that we have to interrupt you, you've applied for the wrong job." And because I'm not shy, I said, "What do you mean, I've applied for the wrong job? I'm a catchment scientist. Australia has been doing this for 30 years. You've never even thought in terms of catchments in this country. You have no idea what catchment science implies. How dare you sit across the table from me and tell me I've applied for the wrong job." They all started laughing at me and said, "No, no, you're misunderstanding. The other vacancy we have is for the chief executive officer. We think you should be applying for the chief executive's role," which is when I started laughing. Seriously, do not put people like me in charge of organizations. The [next] interview for that [CEO position] was hilarious. They offered me the job.

When I landed in the job, I realized that the company was in trouble. Emergency closure meetings every month. Have we got enough money in the bank to run for another month? And another month? I was overwhelmed and felt really ill-equipped. I was really lost.

I don't know where the wisdom came from. It was almost like someone tapped me on the shoulder and said, "Do what you know." And what I know is about connection. What I know is about relationship. What I know is about respect, it's about integrity, it's about value. So that's what I did. I didn't think about business planning. I didn't sit for hours strategizing. I talked hour after hour after hour with hundreds of different people about how they perceived the company. What had the company done, and what might the company do? What would be a value that could enhance the work that they were already doing? How did we avoid competition? How could we position ourselves so that we were only ever complementary, that we will only ever be adding value and never undermining the efforts of others? These were the conversations I had. And lo and behold, ten years later, we're making a modest profit every year. We're attracting large amounts of funding from charitable organizations that gift to environmental companies. And we're very well regarded. My team is just amazing, they're wonderful. And they don't need me, other than to hold the umbrella that protects them from all the bullshit. That's what I do.

When asked to illustrate the kinds of situations she shelters her coworkers from facing, Kat offered the following story.

A couple of winters ago, the British government announced a huge amount of money for tree planting around November. It was world-changing amounts of funding for tree planting. The conditions that needed to be met to secure the money for a new woodland were really easy. They're often very complex, and we keep feeding back to them, "You can't make it this complex. You can't make it this exclusive." This program looked on paper like it was genius. We've had aspirations to put trees in on 50,000 hectares forever. Finally, the money is going to fall out of the sky and we're actually going to be able to incentivize this activity, work with our landowners and get this work done.

The insanity is that it all had to be spent by March 30th, which is the end of the financial year. That's not impossible; you can plant a lot of trees in four months with the right resources. What you can't do is secure the tree stock to put in the ground. Those orders have to be put in the spring of the year before. We couldn't get the trees. We had to go back to the table and say, "Look, we love your work, this is amazing that you've made all this stuff available is absolutely brilliant. You might have just forgotten one little tiny thing." The response was, "That's okay, we'll import them all from Romania." What? They're not thinking about biodiversity or ecosystem integrity, it's just, "Get the trees in the ground." You can't import foreign trees! Why would you do that? Why would you not be insisting on locally endemic species to bolster the local ecosystem resilience? So none of that money made it onto the ground.

If it wasn't for the Deep Adaptation Forum, I don't think I'd still be there holding the umbrella. I think I'd have given up; I'd have been crushed under the weight of it. Because it's really hard. You talk about straddling two worlds. If I hadn't found the second world, the first world would have been my undoing. Because I was exhausted emotionally, physically, spiritually, and mentally. I mean, I am still exhausted. It's horrid. But then I go into a Deep Adaptation space, and I'm reminded of all of the other riches that life has to offer that are away from the nonsense, that system which is so incredibly toxic if you have any sensitivity.

Letting go as freedom

The cost of a thing is the amount of life which is required to be exchanged for it.

Henry David Thoreau

If you could let go of your job, would you? Would you feel freed if you did? If you let go of your car? Of consumer culture? My work is temporary, so it has been easy for me to let go of paid work for a time to explore unpaid vocations and refresh my spirit. All that was required was living on Scott's paycheck, which was easy for us. (Yes, I recognize my privilege. And some of my neighbors manage to be deeply in debt with incomes like ours.)

I did have to get over the message from industrial consumer society that I have worth when I have a paycheck. I am free to work on projects I value that don't pay me. I am also free to step away from the ones that no longer fit and take up new ones that also don't pay, like writing a book for a niche market.

Generous people have experimented with living simply and then produced guidebooks, videos, and blogs. These people are happy to share their experiences, often for free. The Mennonite-influenced *Living More with Less*[85] helped me many years ago to redeem my parents' Great Depression-shaped attitudes toward money. It lifts the value of paying people for their services so they will receive income for their skills and the importance of cooking simply and well for a crowd. Vicki Robin's *Your Money or Your Life* is a classic about freeing yourself from the mental bondage of money and consumer culture.

Jacob Lund Fisker is wonderfully systematic in laying out his approach to "E.R.E.," which could be either Early Retirement Extreme or Emergent Renaissance Ecology. E.R.E. is an answer to the question: "How can I, as a middle-class Westerner, run my life such that I'm not a net drain on the world or society, protect myself from the risks of societal dissolution as much as possible, and maximize my autonomy?"[86] Jacob finds that the work of meaning-making is a crucial and very individual part of stepping away from the rat race of industrial consumer society. He loves learning practical skills to increase his freedom from the need to use money. Remember, he has plenty of time not working at a paid job. Skills he values include sewing, bicycle repair, cooking well, and furniture making. Jacob's generous web resources are fun to peruse as I consider experiments in letting go. He makes theories, tables, and wikis to explain his ideas and experiences[87] because he would like you to join him.

This individual approach is not a fix for our predicament. Nothing is. Instead, it is an opportunity to pull free a few threads of our entanglement in industrial consumer society. As we tug on this tangle of destructive behavior, we discover where the knots are tight or loose. The free play may be different for each of us. The small ways in which we can get free and enjoy our freedom will be valuable information for other people who want to be free from the traps of our industrial consumer society. As we untangle ourselves a little, we create capacity to make small webs of mutual support and skill sharing outside that system.

All this is work: physical work, mental work, and relational work. Or is it? Work is what we must do; play is what we get to do. It does take time and energy to untangle ourselves a little from industrial consumer culture. If you have no time or energy, you are more than tangled. You are strangled. Start by figuring out how to get enough space to consider your options.

Perspective: Maple syrup for oranges and sunshine

Jane lives in Vermont and has been living simply for many years now. She wrote a book about living simply.

In 1986, Jane and her partner Sky made a commitment to home-school and homestead. So they made do on two part-time incomes for a total of $7,000 a year, with two little kids. And they pulled it off. They did that by growing food and taking care of home repairs themselves, not eating out or traveling, and not spending money on much of anything. They lived off the power grid so they had no utility bills. They have never had health insurance, and they didn't get very sick.

Later when their incomes increased, they started saving money by living the same way they had done when they had so little. Despite the additional expenses of moving into town and buying more food instead of growing almost all of it, they still lived very simply.

The family was happy with their own company. They used the library. They still gardened, and they made their own fun with board games and card games. Jane meticulously kept track of all their expenses and income. She knew where every dollar went. She read Vicki Robin's *Your Money or Your Life*.[88] That book taught her to invest her savings until her investment income matched expenses, the "crossover point." She and Sky got to the "crossover point" by saving their modest incomes with this frugal living and were able to retire in their mid-fifties with ongoing expenses of $20,000 a year. In 2006, that was enough for them to go to France and buy a tiny houseboat, spending summers in France and the winters in New Orleans helping to rebuild homes after Hurricane Katrina. Seriously. Please understand: they actually did this on $20,000 a year. Buying a houseboat in France? They got bargain flights and repaired a rickety boat. They rode bikes and took walks. Food and wine were cheap in France. They still didn't buy stuff or go out to eat. Jane has since received two small inheritances that she has used to build and rehab modest homes. Jane is not worried about economic disasters affecting her. And she is generous with gifts and hospitality. She *will* spend money on one nonessential that she loves: travel. But she doesn't spend much.

Once, when her children were little, and her parents lived in Florida, she had no money in the budget for travel. Determined to see family and sunshine, she popped the kids in the car and drove from a Vermont winter to the Florida sun. She funded that car trip by bartering twenty gallons of her homemade maple syrup to a Florida orange grower and then selling all the oranges her car would hold back in Vermont.

Jane has never owned a dishwasher, microwave, dryer, or any fancy home gadgets. She claims that's a significant way to save money. She swaps and borrows with a neighborhood buy-nothing social media group, and she improvises. She invites us to try going a week or a month without visiting the grocery store. It might not be fun, but it would be a learning experience. We can build confidence to face future shortages. More guidance from her experience is found in her book *Freedom through Frugality*.[89]

Jane can live simply, and enjoy it. So I can probably live with much less than I think I can, and enjoy it. Jane loves the freedom and confidence her lifestyle has given her. And she has a large plot of land by a lake in Vermont where you can build a tiny home and garden with her.[90]

Many voices: Simple living

Here are specific things people have done to live more simply that they found fun, rewarding, illuminating, or satisfying. I have done very few of these things, but they expand my ideas of what is possible. I was overwhelmed with responses in the Deep Adaptation private Facebook group on this topic. This is a sampling:

- Use it up, wear it out, make it do, or do without!

- I think doing anything is better than nothing. Our culture is fostering our uselessness so that everything can be sold to us. I think doing anything to push back against the de-skilling is good. And from an adaptation perspective, I think that coming to know that you can do things helps you try new things. We are all going to have to do a lot of things we are unfamiliar with in the future, and not feeling incapable is a great start.

- What makes you feel nourished? I left my university job. I work as a community organizer these days. We run mutual aid food hubs, support local farmers markets, do antiracism work, and tend to the permaculture gardens. It's what makes me feel nourished.

- I invite folks to take up a meditation practice aimed at helping to regulate their nervous system… Developing… strategies to manage stress, and lowering my stress more generally, made it easy to let things go without having to deal with the withdrawal symptoms.

Cooking and gardening are popular:

- I learned how to walk into a kitchen and cook something I liked out of the ingredients or the leftovers [at hand]. My recommendation is to focus on 3-4 dishes you really like (pizza? soup?) and learn how to do them well. If you specialize in a dish, you figure out how to optimize it for your taste. Practice makes perfect! Warning: Going out to eat will subsequently feel like a waste of money.

- I am co-gardening with a neighbor instead of belonging to a community garden that I was driving to. We are planning next year's garden together and will be digging up a chunk of her sizable lawn. [Renters take note!]

- I have been harvesting a lot of annual and perennial 'weeds' that grow abundantly around here. I didn't even plant spinach this year. We get lambs quarters, chickweed, redroot pigweed, nettles, sow thistle, and plantain to name a few. They are great in salads or stir-fries. I even blanch and freeze them.

- For us, growing as much of our food as possible has been both freeing and fun. And so much more healthy. It means cooking and eating at home. We share with friends and neighbors. We work to close the loop, so compost and use our urine for fertilizer. I have stopped buying any outside amendments. It's very rewarding to learn new, simple skills.

Various challenges and systems can kick-start a new lifestyle:

- Many years ago I did a 30-day minimalist challenge, and it completely changed how I viewed 'stuff'. It opened the door for a lot of changes in a pretty small amount of time.

- I usually invite people to do a "buy nothing" challenge for a minimum of 9-12 months. It forces one to become creative and find meaning outside the consumer system when it's no longer possible to buy solutions… The first three months are pure withdrawal and "complainy-pants" mode. After half a year, people have found viable substitutes. After 9-12 months, some don't want to go back to their old ways.

Some other contributions:

- I got rid of the TV 20+ years ago. Gave my car away a couple of years ago to a young man who needed it to get to work. Now I walk almost everywhere, bank, grocer, pharmacy, post office. For medical trips, my area does have an option if I call a week ahead.

- We have a group of people we trust, and they do not necessarily know each other, but if today I have extra produce (I have enough canned, fermented, and dried) then I just give it to someone who can use it. I trust that they'll reciprocate at some point. It's been such a beautiful thing! It's happened with all kinds of tools, random household items, and labor.

- I enjoy sourcing lovely mix-and-match clothes from second-hand shops; creating new and interesting recipes using garden produce and herbs and local veggies; potluck lunches with friends rather than eating out; putting a pretty sticker over a slight dent in the car, rather than paying to have it removed; paying to go on a course then sharing the info with friends who give a donation towards the cost of the course and friends doing the same; sharing garden produce, seeds and surplus plants.

Simplifying entertainment:

- We've been running a living room and backyard open mic for years, anyone can have the floor for 5 minutes to read, recite, do magic tricks, play an instrument, sing a song… No technology allowed, like in generations past when people gathered in the evening and entertained themselves and each other. Bonuses… long hangouts with great conversations after the amateur hour ends, and BYO is a lot more affordable than a night on the town.

- In the US so much of our culture is wrapped around going out to eat or for drinks. But having a picnic (just to throw an example out there) is so much more leisurely and fun.

- I really like our song circle. We have a leader and she teaches us the songs each session. They are usually short and simple; we can learn them in one go. And then I try to sing some of them at home, instead of using other media for entertainment.

- Board games– social, communal, build strategic thinking, zero emissions other than from manufacturing (and a well-made and cared-for game can last for generations).

Simplifying housing:

- We do all the homesteading activities, plus we're off-grid. We do lots of up-cycling, especially making clothes from other textiles, and making rugs from rags. Our home is passive solar, so almost no heat is required all winter, and no AC in summer. I am cooking with a solar oven and heating water with a solar breadbox hot water heater I built. This puts us on nature's clock—no sun today, I don't do dishes. Maybe that's just me being lazy.

- We've just traded houses with our daughter, son-in-law, and 4 kids. They get the big old 5-bedroom, 4-bath farmhouse and we get the 920 square foot mobile with 2 bedrooms and 1 bath. I'm learning how to live in a small house. It's very liberating. Why do we need 12 mugs and 6 cups and saucers? I'm tossing and donating so much stuff!

- Our house is much smaller than the houses of our peers, and in much worse cosmetic condition. Everything is crooked. It has mice so we got a cat, and it is much more crowded compared to our peers since we have 4 kids. We love it. This has allowed my husband and I to limit work hours and set clear boundaries around how much of us our workplaces can have.

Changing our relationship with money:

- Having no debt is a HUGE simplification. I recently opted—when my laptop died—to get a refurbished computer at a price I could pay off in one month, rather than getting a new laptop that would have dipped into my savings or caused me to use my one credit card. Not only do I stay out of debt, but I am recycling technology. Double win.

- Before I buy anything, I ask myself why I want it. If the answer seems reasonable, I then ask myself if there is another way to get what I want without buying new (can I borrow this item? can I find it used? can I substitute something else for it?) I've been at this a long time, so the conversations with myself tend to be rather short at this point.

What approach toward living more simply appeals to you? Remember, the point is to experiment, learn, have fun, and be free to make choices that fit your life and your values.

Summary and reflection

- Security is an illusion. We are all vulnerable. We always were. It's best to face that reality and practice letting go of the extras of industrial consumer society that will not last.

- Losing her home to wildfire tested Carla's Buddhist teaching about facing impermanence. No matter how sound our philosophy is, profound loss still affects us bodily.

- A vision of what might happen when times get tough. Dancing helps.

- A ritual for letting go: fun, and suitable for a group. Try it!

- Letting go of the extras of industrial consumer society is solidarity with the wider human and nonhuman world. Let's have fun experimenting with living more simply.

- Have a Plan B for when Plan A doesn't work out.

- Letting go is an excellent skill to practice when living with others.

- Kat describes her quest to keep contributing her skills to people and to our Earth home. It has involved a lot of letting go.

- Letting go can be liberating: from the rat race, from stuff, from bills, from financial enslavement.

- Jane shared a little of her experience with living simply over many years, including a crazy road trip.

- People share all kinds of ways they live simply, at least by the standards of our industrial consumer society.

CHAPTER TEN

No more flying solo

Relationships are the basis of resilience in hard times. In this chapter, you will find reflections and experiences that explore the power of community and partnership.

Connecting across differences. Challenging ourselves to connect beyond people who are like us.

No more flying solo. My experience of learning to work with partners.

Being neighborly. You need not become intimate or have much in common.

Perspective: Swaps for building community. Wendy experiments with being neighborly in rural Portugal.

Perspective: I left my fortress. Raewyn moved to a more urban area for community.

Community for cultivating influence. Political influence is about relationships.

Perspective: At least one radical leftist prepper. Daniel believes preppers shouldn't fly solo.

Belonging beyond the human community. This approach is worthwhile, whether or not you think it's possible.

Kindred spirits. Supporting each other in common goals.

Connecting across differences

When seeking human connection, the easiest path is to find people who think and act like us, share values and culture with us, and understand our jokes and our slang. If we insist on finding just the right people to live and work with, though, we may be setting ourselves up to be lonely indeed. A little group that shares our values is a precious gift, but we need not surround ourselves with those people. A regular check-in may be enough.

Industrial consumer society disconnects us from each other by making so many interactions transactional or competitive. Computers, the internet, and cell phones substitute more and more for actual human contact. COVID made our isolation that much worse. So we are starting at a significant disadvantage.

Looking for that particular person or group where you fit in harmoniously is fine, and you may never find them. If you do, it likely won't last. Where I live, an action as simple and profound as finding a romantic partner is machine-mediated by dating apps. How is it working? At this writing, all but one of the major dating apps in the U.S. are owned by the same company,[91] whose business success depends on unsuccessful lovers coming back for another try. Maybe the relationship that will most nurture you, or the community where you can most learn and contribute, is one where you didn't predict a good match. Maybe feeling awkward sometimes is part of being in relationship. When not all of your views are shared, you must search together for common ground and build bridges.

When you know and are living your values, you may find yourself withdrawing from certain groups because the clash of differing values is so uncomfortable. If you are not physically safe, withdrawal is probably your best option. If you are safe, consider the following: are they hard to escape? (family, neighbors) Or do they share some significant bond with you? (coworkers, long-time friends, people brought together by some common interest) If so, learning to affirm the values you do share, even if that's only eating pie together, while gently deflecting or questioning those values that trouble you, is a great skill to build. Respectful disagreement is a crucial skill for collaboration, community building, and preventing violence.

No more flying solo

I came to the online Deep Adaptation Forum (DAF) to offer my "Grief Gratitude and Courage" workshops. Master facilitator Katie Carr advised me to participate in others' workshops and to always have a cohost for my own. That was the request of all facilitators in DAF. A Zoom cohost made sense, especially with all the moving parts in my workshop. But as I spent more time hosting groups and cohosting others' groups, I began to appreciate how much more was possible with a good team. My facilitation skills grew as I witnessed others in action, and sweet synergy occasionally made the experience amazing for me. Some fine friendships have grown out of that cohosting.

When I first started participating in facilitated events in DAF, I was surprised to see many other facilitators as participants. Sometimes it seemed like we facilitators were mostly facilitating for each other. I had in my head that facilitators were the providers and participants were the clients. What I came to realize is that all of us benefit from being supported in DAF spaces as well as providing that support. It is a form of mutual aid. In Deep Listening (or other circling approaches) the listeners and the speaker are all contributing. I am so grateful I don't have to face collapse alone.

Nenad, who has been the Deep Adaptation Forum's network weaver, taught me the power of project workgroups. So when I took on a big project, I gathered a workgroup. I did most of the work, but I had neither the authority nor the organizational skills to do the project alone. I met weekly with the team. One thing I learned from that experience: I will want to convene more people than I think I need. Then I can find out how each person will truly enjoy contributing and empower them to do that.

I have a partner, Lisa, for events administration. She and I work together easily. We have different strengths, so we are a good team. Nenad says at least three people are needed for a sustainable work group, and it's easy to see why. But two are a whole lot better than one. He also says that spending time on personal connection and fun is essential when meeting. That is true in my experience. I appreciate the suggested framework of DAF meetings: Take time for a generous check-in, and if a check-in reveals something big in someone's life, we attend to that. We are people in relationships, not machines in production.

These are small ways I am healing from the brokenness of modernity. I don't fly solo as much anymore.

Being neighborly

Community, I am beginning to understand, is made through a skill I have never learned or valued: the ability to pass the time with people you do not and will not know well, talking about nothing in particular, just to be sure of each other, just to be neighborly. A community is not something you have, like a camcorder or a breakfast nook. No, it is something you do. And you have to do it all the time.

– Wendell Berry

Knowing and trusting our neighbors will equip us to face whatever happens better than we could alone. Neighbors are not always easy to get to know and trust. Yet we will have to rely on neighbors if transportation systems are disrupted. They will face the same challenges as we encounter, and pooling our resources will equip all of us to weather disasters better.

Irv Mills writes a blog called "The Easiest Person to Fool."[92] He offers these reflections on building community with neighbors:

> To succeed, community-building efforts must be based on clear and present needs. If you're living in an area where collapse has not yet struck, where business as usual is still "working" fairly well, then trying to put a community together because you think it will be needed someday isn't going to work. Those involved (including you) simply won't have the motivation to make it work, to stay together, when there are easier alternatives all around you. Especially when that community is made up of people who haven't yet had much practice at such things. A quick look at the history of intentional communities will show how hard it is to succeed at this.
>
> So first, take every opportunity to work, and play, with people in your community. Build a network of friends and acquaintances. Get a reputation for contributing, reciprocating, and carrying your weight. Then, when the need arises, you can get together with people you already know and respond more effectively.

It's unlikely that you will have neighbors who share your perspective about unfolding collapse, so it's best to let that topic go. Instead, work to get to know them. You could bring them baked goods. That's my favorite approach. You could invite them over for a beer. That's my close second. You could ask to borrow a ladder or a lemon. Asking to receive a small favor from a neighbor rather than giving one might feel awkward. That's the point. That minor imposition makes it more comfortable for your neighbor to ask you for a favor too.

One December, I knocked on all forty doors in my court, taped up invitations for the people who weren't home, and invited them all to a holiday open house (pre-COVID). At the party, I collected contact information from those who were willing. It sits gathering dust, but in case of emergency, it exists.

If you have the energy, you can start conversations about emergency preparedness. Couch it in terms of preparing for situations that are recognized risks in your area. That could be as simple as electrical outages. In my area, it includes fires and earthquakes.

Jessica Canham speaks of the people in her community on the small Caribbean island of Dominica, most of whom are subsistence farmers: "Volunteering is second nature for people. We don't even call it volunteering. Service work is something everybody does. Everybody puts in the time and the energy because you are always on the receiving end as well. And you never know when you might need more support."[93]

Most of us are far from this kind of relationship with our neighbors, but we can make a start.

Perspective: Swaps for building community

Wendy is South African and has lived in India and England. She wanted to cultivate a garden, to live close to the land, so she bought a small house with a bit of land in rural Portugal. She wants to cultivate community with her neighbors too.

Many of the people who historically lived in Wendy's area of Portugal have already left for the more prosperous cities. British, German, and Dutch pensioners have moved there for the sunshine. Their pensions make them wealthy in comparison to most of the locals. Most small farms still belonging to locals have been in their families for generations. Some families have sold off most of their land but retain the house and a generous smallholding for a garden. Those locals who remain grow an impressive amount of food and make do by fixing and improvising; some have jobs nearby.

Everyone in the region was deeply impacted by horrific forest fires in 2016 and 2017, in which a number of people died. Those fires, and the 2022 drought, have brought home to almost everyone the fragility of life in this area, whether or not they are collapse-aware.

By building community and finding common ground for the locals and the immigrants, Wendy hoped to create a kind of "resilience insurance" where neighbors are sharing and learning from one another. She has long experience as a yoga teacher. So she offered community yoga classes on a donation basis. Donations went to the local fire brigade and a dog rescue shelter. She found that the idea of donating back to the community was a draw. The group got big fast; they had to limit enrollment. They met at least every week for a couple of years. From yoga came dinners, craft fairs, and more.

Wendy had noticed that small farmers and gardeners always seem to have too many tomatoes and not enough cucumbers or the other way around. So more recently, the group has had great fun doing something that comes out of the Transition movement:[94] produce swaps. Wendy told everyone to share the invitation on social media. Anyone could come; you didn't have to bring produce to swap. She served teas and coffees, and that offering inspired neighbors to bring a formidable feast of cakes and savories. To the delight of all, it became a big party, with people sharing tea and treats and the swap going on in the background. People then wanted to have a swap every week.

These swaps are more than fun. Referrals are given: this person knows the trick of growing this or how to fix that. These connections are of great value to people trying to live on the land for the first time in a very different climate from their old home.

A woman in a town a bit too far away from Wendy wanted to start her own swaps, and she is a force of nature. Old Portuguese villages typically have a weekly market in the central square that nowadays is a commercial market; items for sale are mostly not local. This woman convinced the mayor of her village (president, he's called in Portuguese) to let her use that space, and now her swap is one Sunday a month in the village square. In addition to an informal social gathering, traditional dancers perform, and who knows what will happen next? This community-building woman is Dutch. She now works comfortably with the Portuguese town leaders. Progress indeed.

Wendy was hesitant to resume swaps after the COVID lockdowns. Another neighbor was begging her to restart and finally said, "Can I just try hosting one?" Wendy knew better than to say

"No." So this woman invited people to her enormous, luxurious backyard garden. They gathered under two venerable oak trees. The host had a quarter acre of vegetables, flowers, perennials, and fruit trees that people were drooling over. She got loads of admiration and encouragement for this nine-year-old garden and will now be consulting for her neighbors' gardens.

The swap aspect has confused some people. They might be hesitant to take something if they come empty-handed, even if it's from an overflowing mound of vegetables. Or they may think a valuable item should command a price. Wendy and the other organizers hold a firm line on this. They want people to be creative in figuring out how to work exchanges that don't need cash. Sometimes they sit down and explain that money is a mutually agreed upon token and that other ways of exchanging are also valid.

Wendy wants us to know that she had no specific plans or goals that led to these results. The events unfolded organically, as she was willing to take the next step and organize more of what people found exciting. Then she could try a new angle and see how her community responded. Another neighbor wants to start a local "buy nothing" group on WhatsApp. And the neighbors keep offering to help. Now that's community!

Perspective: I left my fortress

Raewyn lives near Wollongong, Australia, at the seacoast about 90 kilometers south of Sydney. She recently made a counterintuitive move from a semi-rural home in a far outer suburb of Sydney to this more densely settled area as a way to find community and to live lighter on the land.

About 15 years earlier, Raewyn left Sydney for the Blue Mountains, an hour's drive to the west. She had dreamed of a spacious garden. She imagined becoming self-sufficient. She wanted to be closer to the land. And she did get to garden. She also installed insulation, solar panels, and a rainwater catchment system. But the home she had was never comfortable. No matter how she tried to retrofit it, it was drafty and cold in winter. And it was too much space for one person.

Looking back, she thinks her move was not the wisest. The steep slope on her property meant she could only cultivate a small garden on her large lot. An extensive property and a poorly-designed home were challenging to maintain. They required more resources than she wanted to use. She felt the irony when she took a job back in Sydney, driving that long commute regularly because train service was limited. She and her friends sometimes worked together in their gardens, but people were mostly isolated day to day. Raewyn craved some communal living and working and playing. She envisioned preparing for coming hardship with a community, like a swarm of bees.

For these and other reasons, including the present danger of fire, she decided not long ago to move to a more densely populated area near the coast. With her new partner, she chose an apartment in a block of 20 units. Their apartment's association, full of disagreements and personality clashes, is now part of her communal living. She finds this amusing. She downsized her belongings and lives more lightly on the land by having a smaller space and shared walls, which pleases her. Raewyn spends little time caring for her apartment home, so she can spend more time working in a community garden, on bush lands restoration projects, and doing other community projects. She enjoys working in community gardens rather than maintaining her own land, going "from ego to eco," a phrase she learned from Otto Sharmer's ULab. She appreciates the free-ranging conversations she has while gardening with people. She has a list of community projects a mile long for when she retires. "I left my fortress and came to where I thought life would be more communal." It seems to have been a good move.

Community for cultivating influence

Protest that endures, I think, is moved by a hope far more modest than that of public success: namely, the hope of preserving qualities in one's own heart and spirit that would be destroyed by acquiescence.

Wendell Berry

Cultivating influence is just knowing people and building connections. Reading the news, talking with friends and family, and posting on social media do not cultivate much influence. I am considered an influential person in my community, which amazes me. I know a few people who make things happen in politics. I know more people who know people who make things happen. How do I know them? I worship at the activist church. I go to political rallies, and I knock on doors to help candidates get out the vote. I am an active member of several community organizations.

I make friends with people who get things done. This isn't hard, and it isn't self-serving. We care about the same things; we support and care for each other. Despite living in a city of 300,000 in a county of 3 million, the networks of people in the know are not so large. By being of service to a couple of organizations and taking the time to make connections, I know people who know people, and that is enough.

This doesn't mean I get what I want in politics. It means I usually hear about actions being taken and the responses to those actions. I know when to best use my allotted 3 minutes of speaking time at City Council to influence a proposed ordinance. I know who to ask about what happens to our recycling. And I support others in small ways, including helping immigrants and refugees navigate the labyrinth of unjust laws and procedures that separate people from their families and disrupt their lives. I see my role more as a witness and truth-teller than change maker.

Influence can mean knowing who to call and where the resources are. Influence can mean advocacy and mutual aid. When crises come, I will know who to call, and who knows where the resources are. In the meantime, I know who to talk to for inspiration and who to invite to make a party lively.

Perspective: At least one radical leftist prepper

We met Daniel in chapter two. He lives in suburban Southern California.

Daniel began his journey of collapse awareness with concern about U.S. societal breakdown as he watched the pandemic drag on and on. He was shaken by the murder of George Floyd and other murders of people of color in the U.S. in the summer of 2020. He began reading about the state of democracy. Disturbed by what he found, he kept reading and discovered the state of the planet. Soon he was spending all his spare time reading about ecological, political, and social collapse. He got a sense of the root causes and the magnitude of the predicament.

There is always more to learn, but at some point, he asked himself, "What can be done about it?" Not necessarily to prevent things from collapsing, although that was his initial thought. One thing he's doing breaks stereotypes. He's prepping. He was resistant at first because he had always assumed, "Prepping was a right-wing subculture of people who build bunkers and stockpile their AR-15s and wait for the zombie apocalypse." He has been pleasantly surprised to discover that the preppers he has met are intelligent, reasonable people with various political views. These preppers welcomed all the newcomers joining them because of COVID; they felt validated by the attention.

Daniel thinks prepping can be for the good of a community, laying in a stock of food and water, and other emergency supplies. He likes to think of prepping as a way to "Put on your own oxygen mask first," so you can help others. He believes prepping for selfish purposes doesn't work anyway. The lone homesteading family could, for example, face appendicitis, tooth infections, or other challenges that require a wide range of skills no single family can have. He believes we will need community in times of collapse. He's still trying to figure out how to create that kind of community in the faceless suburbs of Southern California.

Belonging beyond the human community

Modernity tells us that if we seek support in dealing with hardships through deep and personal relationships with the nonhuman world, we might have mental issues, or just be flaky. Yet in all times and places, people have been doing just that. You may not wish to seek your support in the nonhuman world, but please know this support is legitimate and valuable to those who do. More people rely on this kind of support than talk about it. No wonder, since meaningful relationships with the nonhuman world have not been recognized as valid in my culture. Many scientists have a deep reciprocal relationship with nonhuman entities. Cultivating such support is becoming a little more acceptable in the mainstream. Practices such as forest bathing, tree-hugging, and walks in nature or urban landscapes with a little green are now recognized scientifically for their value in reducing anxiety and regulating emotions. Most traditional cultures have language and practices that support these kinds of relationships. Traditional and nontraditional practices that honor the seasons, ancestors, nonhuman entities, and spirits are also ways to receive support beyond flesh-and-blood humans. You can experiment with these relationships if you choose.

Some traditional teachers would tell us that these nonhuman entities are demanding taskmasters and that they are angry at our disconnection and disrespect of the nonhuman world. This would hardly be surprising. I have heard elaborate directions for engaging nonhuman entities, demanding large time commitments, with warnings that you must get it right. Yet, in my experience and that of many friends, an intentional search for compassionate helping entities almost always turns up wise, kind, and nurturing communications. If a human teacher or method leads you otherwise, please disengage from that person or practice for your safety. Find support in cultivating nonhuman relationships elsewhere.

Some people talk to trees. Sometimes the trees talk back. Others just feel better after a half hour of sitting under a beloved tree. Both experiences are valid. In relationships beyond the human, we can draw no clear line that has "real" reciprocal experiences on one side and the workings of our own minds on the other side. Therefore, you can come with the expectation that entities beyond the human exist who want to partner with you for your well-being. I trust this is always true if you want to be of service to others, human or nonhuman. Hold your experiences lightly if they are hurtful or confusing. But if your experiences with the nonhuman world are supportive and empowering, claim them. You will be in good company, both human and nonhuman.

Kindred spirits

"Sorry, I'm not very organized," said one member of our little Deep Adaptation Resilience Accountability Group.

"Don't worry," I replied. "None of us are very organized. That's why we're here." We meet on Zoom for an hour every other week to support each other in making practical changes in our lives. We often don't know what to do, or if we do, we don't know how. We aren't very organized or efficient. And life gets in the way sometimes.

We take time at the beginning of each meeting to check in. "How are you, really?" We sometimes amuse ourselves by calling it "the weather report." We take time to acknowledge any unusual or challenging conditions, both outside our doors during a record-breaking summer (oh wait, aren't they all now?) and in our internal weather, our mood and health. We already know that a key part of showing up for hard times is listening to one another and offering a supportive presence. Rough weather, inside or out, is easier to bear when you are not alone, when you are heard and not dismissed. We are kindred spirits, and being together with a common outlook and goal, whether or not we're progressing quickly, empowers each of us.

I treasure my kindred spirits. Some of them are in Vietnam, Scotland, Vermont, British Columbia, and Idaho, so that means Zoom calls. They share my passions, and they understand my motivations. I learn from them and respect them. When we talk shop, we naturally affirm and inspire one another. We learn from each other without working at it. It feels like play as the meeting time flies by. Because we are vulnerable in our passion, it is easy for us to get personal and share personal things beyond the stated common ground.

Kindred spirits do not have to match all our passions. The people fighting for immigrant rights with me don't need me to announce the staggering numbers of immigrants I expect us to see in the next few years fleeing climate chaos and its fallout. These friends are doing the hard emotional work of caring and compassionate witnessing for desperate, hurting people now. They will be able to transition when they are needed. Enough for now that they do what they do, and I get to learn from them.

I sometimes share my dark vision of the future with people in leadership positions or with close friends. I don't expect them to agree with me, just to understand me a little better. I am grateful that they take me seriously as I share my perspective. Talking about this book has been an interesting entry into that conversation.

Where does one look for kindred spirits? You can try groups that tap into your interest or spark your curiosity. For me, those have been activist groups, spiritual and educational programs, and peer groups for professions, vocations, and hobbies. If you put yourself out there, you will find people who share your passions.

I want to be careful to see those kindred spirits as whole selves, not just as co-laborers for a specific vision. This change of perspective takes unlearning and relearning on my part because it is not efficient. Instead, it is resilient. It connects and strengthens us in community, things I was not taught to value and prioritize. I am learning.

Summary and reflection

- Finding community is challenging in industrial consumer society. But it's worth working for.

- Don't expect or settle for only comfortable relationships. Maybe the relationship that will most nurture you, or the community where you can most learn and contribute, is one where you feel awkward sometimes. When not all of your views are shared, you must search together for common ground and build bridges.

- I have learned the power of partnership in my work. I no longer want to go it alone.

- Befriending neighbors does not mean becoming intimate with them or agreeing with their views. Instead, it means getting to know them, borrowing, sharing, and having get-togethers. In times of trouble, neighbors are the people we will turn to first.

- Wendy has been plotting fun activities to help build community in her neighborhood in rural Portugal. Too many tomatoes are an excellent excuse for a tea party.

- Raewyn moved to a more urban area in Australia to find more community.

- If you want to influence local politics and government policies, get to know people. Contribute, and make connections among other people who contribute. Those are the people you'll want to know in an emergency.

- Daniel found community in an unlikely group, preppers who have their passion for disaster preparedness as common ground.

- We can find community and support in the nonhuman world too. Though the modern world has little room for this traditional practice, people do it anyway.

- Finding kindred spirits and meeting with them for mutual support and collaboration has been a reliable way for me to cultivate dear friends. We attend to each other, not just the work at hand.

CHAPTER ELEVEN

Young people and those who care about them

Climate anxiety is taking a toll on children and young people around the globe.[95] They see the evidence before them with clear eyes rather than the practiced denial of their elders. In this chapter, I offer some perspectives by and for young people, starting with the voices of three young adults. Britt Wray has also written a valuable book from the perspective of young adults.[96]

Perspective: Making a life in strange times. Ellen's goal is a happy life.

Perspective: Grim, but not resigned. Andrew fears for our political future, so he is politically active.

Perspective: Supported and supporting. Ari supports climate activists younger than she is.

Telling children the truth. We can do so in age-appropriate ways while doing our own inner work.

Providing children a secure base. This looks different at different ages but is the foundational gift we can give children.

Perspective: Teaching children ecophilia. Peter teaches children to recognize and love the nonhuman world around them in his forest school.

Many voices: What do the old owe the young? In our speech and our actions.

Being responsible elders. Some ideas to address this lack in my culture.

Perspective: Making a life in strange times

Ellen lives with her parents in northern Scotland. Her mother is a climate scientist, so she grew up with an awareness of the dire state of the planet. At 23, she recently graduated from university with an art degree.

Facing an uncertain future, Ellen doesn't have a plan, and she doesn't want a plan. She claims she is taking after her mother by winging it. "I think my one goal, which is probably a really good one to have in a time of climate crisis, is I want to be happy. Everything else is negotiable."

Art was her major in college. She does art now because it brings her all kinds of blessings and it takes her away from the world's troubles for a while. She notes that art can also help people express those troubles. She can produce custom art to sell, pet portraits to be exact, and that's not a fit for a day job. For now, she chooses to work in a grocery store called "The Co-op." While still somewhat corporate, it allows members to vote on things like raises for her and her coworkers. It also allows her to make a modest contribution to an organization that helps children experience the outdoors. Recently she made a fuss about the store selling frozen Asian shrimp when shrimp are being caught 20 miles away.

Ellen does feel mood effects from the state of the world. "It's something that I've got to come to terms with and not let it crush my spirit." She offers an unusual strategy for helping her mood. "I have a massive abundance of houseplants, probably 30. I got my first house plant in university from one of my girlfriends. She got a load of houseplants to help her cope with her anxiety and depression. You see, plants can sense those kinds of things, and they will not thrive in an unhappy environment. I have these living things that need my care and attention. It has helped."

She notes that even her friends with lucrative jobs are realizing that they'll probably never be able to afford to buy a house. She is tired of people with paid-off mortgages telling her generation to "work harder." She lives in a rural area with her parents, who enjoy her company.

Her mother was always honest when Ellen asked difficult questions growing up, even when the truth was messy. So she was never under any illusions, and seeing her mother cope helped her handle her own worries about the world. At university, when she brought up topics of climate collapse or social collapse, asking what might be done to prevent the human race from going extinct, people would call her crazy. A couple of close friends understood; together they shopped at the local greengrocer, finding awkward-shaped cucumbers and enormous carrots and enjoying not buying corporate. At a recent music festival, though, she heard people who were aware, who were not expecting to go back the following year because they can no longer afford it and other things were a priority, like learning to grow their own food even in little apartments. She wonders about the unraveling that the future will bring, but she trusts that community will matter, growing food will matter, and finding ways to exchange without money will help too.

With a boyfriend 500 miles away, Ellen finds herself on buses and trains for long periods. She has discovered a way to feel useful on those journeys: she takes an anti-racism book with her. People walk past, see the cover, and either smile at her or turn away. That way, she knows who to talk to. A young man came and sat next to her on the bus. Sitting down, he turned to her and said, "I figured the girl reading that book is my best bet to have a safe journey home as a black man." She was happy to have taken that book with her. It meant that he had at least a short period of his journey where he knew he was supported. She thinks every journey should be a safe journey.

Perspective: Grim, but not resigned

Andrew is a 29-year-old professional musician who grew up in Southern California and recently moved to the more affordable city of Rochester, New York. He became politically active after the election of President Trump. He went from volunteering to being a paid worker on campaigns of people whose platforms he supports. I asked him to speak about his experience of the state of the world, and the perspectives of friends and colleagues around his age.

Andrew confesses to being a worrier. We talked about his worries about the future of democracy and economic security. He recognizes that addressing climate change through the political process is hard since, as he put it, "The cat is out of the bag," already. Many of his friends take an attitude of grim resignation toward that catastrophe. But he is not passive; he engages in the political process not just out of duty but to make a difference.

He is frustrated about our political failure to address income inequality in a state where the more liberal party has been in power for many years. Andrew only knows one person his age who has managed to buy a small condo in our area, with a loan from his grandparents in addition to hard work and savings. Some of Andrew's friends have moved away to places where housing is more affordable, and he recently moved as well. He said, "As long I've been paying attention to politics, I've heard, 'We need more affordable housing, we need to fix housing prices, we need to do this…' and it's just gotten worse." He is also disturbed by the level of infighting he sees among local progressive politicians and activists. "Everybody is a single-issue person," he laments. He suspects the political right is more unified and better organized, and that is not a comforting thought.

He is troubled by racialized police killings. He was enjoying a visit to the zoo until he noticed all the signs saying that the animals he was visiting were in danger of extinction. He has the sense that when he was growing up, he was told how the world works. But he has found out that so much of it is B.S. He sees many people his age doing their best to ignore the troubles that trouble him. As the waves of bad news hit, these young people become ever more disillusioned. He describes it as, "A massive feedback loop, which ends up in grim resignation. Just put your head down. Do what you can; try not to think too hard, you know?" He is cynical about the advice he was given in school to "follow your dream." For musicians, that dream was always difficult, but now it seems impossible.

Andrew's preferred methods of coping are:

- Doing something political. He feels a responsibility to work toward the world he wants. Whether or not there's any hope of his work succeeding, just doing something helps.

- Choosing to separate himself from people who drift far from him on the political spectrum, to the point of conspiracy theories. He doesn't want to add to existing polarization, but sometimes he's just had enough.

- Dark humor helps. It comes out in small groups where people feel safe to express those darker views. He thinks his friends' satire is sharper than it used to be.

- Like many people his age, he's done some therapy. He sees people being more open in sharing their mental health struggles than when he was a teen, and he thinks that's a positive step.

- Though it's not his thing, he sees that religion can be a huge help in coping with life's dark side for believers.

Andrew doesn't see a separation between nature and the human environment. He explained, "The coolest kinds of green-themed stuff that I've seen, that really get people excited, are gardens in cities and green spaces in cities. How do we integrate more nature into our spaces? So there's less of a divide between the city and nature." He sees a recognition among his friends that we need to address the whole; there is no nature in isolation.

Perspective: Supporting and supported

Ari lives in Southern California. She is 29 and is finishing up a Ph.D. in Environmental Engineering, working on flood prevention. She is a mentor for young climate activists. She helped start the Los Angeles chapter of the Climate Reality Project. Her latest project is "Reform and Sustain," a group for young climate activists from middle school through graduate school.

Ari knows that a certain amount of climate change disaster is inevitable. But she believes that more is avoidable if a critical mass of people take action. She wants to be a part of making that happen. She told me, "I believe action is the antidote to despair."

Ari has found a day job that supports her goals. She is studying how to prevent flooding in different communities in the aftermath of fires. Flooding doesn't always conform to the models that engineers use. Flooding patterns in a wealthy coastal California suburb look very different from floods in Tijuana, just over the border in Mexico. Because her research group, the Flood Lab at the University of California at Irvine, took the trouble to talk to people in Tijuana who had lived through floods, they learned how important it is to revise their models to reflect different affected communities' situations.

An elder has mentored and supported Ari in the Climate Reality Leadership Corps training. And she, in turn, is supporting people younger than she is. She wants the group "Reform and Sustain" to be a place where young people can bring their concerns and anxieties and hold space for each other to talk and be listened to, which they seem to need and appreciate. She trusts that the young activists she supports are capable of talking about challenging topics that many adults wouldn't bring up with them. When we spoke, the group had just discussed composting burials. That topic was their idea, not hers.

These young people can feel isolated in their daily lives, with few friends to share their passion. They already know that activism takes a lot of effort and can take an emotional toll. "Reform and Sustain" has social gatherings where they eat together, something she doesn't see much in her activist groups with older members. This leaves them feeling connected and committed to the group.

In her hectic life, Ari stays grounded with a daily Transcendental Meditation practice. TM is a simple style of meditation that uses the repetition of a mantra. She has had spells of grieving. She explains, "When I heard that we were pulling out of the Paris Agreement, I got upset because I thought, 'How many people are going to needlessly die because of this?'" What gave her the most hope during that time was finding other people who cared, who thought they could make a difference, and with whom she could work.

Ari shared why she does what she does. "What motivates me to keep doing that is I know there's a possibility that I will have a future where my kids, if I decide to have kids, or at least future generations... I can tell them I really tried to do the right thing. I really tried to make the world a wonderful place for you."

Telling children the truth

When the elders tell the truth, the young people will be fine.

– Owl Drums

Just tell children the truth. That sounds simple, right? But of course, it's not. Isn't the truth too dark? That is a challenge that we need to overcome for ourselves. We need to know and be able to tell what we understand of unfolding troubles and to admit what we don't know.

Younger children don't need to know. Notice what your child fears and tread lightly. Teach them tools for relieving fears, without expecting immediate results. Notice how they find strength and courage. Lift those skills up for them. Allow them to dream. And don't be afraid to admit what you don't know. "I don't know what will happen. But we'll get through it together. I'll be here for you."

Tell children stories of courage and resourcefulness and generosity. Tell them about your heroes and mentors, the people whose actions embody your values. Tell them stories of hard times and how people got through those hard times. It's a great excuse to discover and learn those stories ourselves.

As children get older, we can do our best to counter prevailing messages, such as: your worth is found in money, status, and possessions, if you don't have those things you are a failure, or the worst one —you are responsible for saving the world.

Knowing that we don't know how to live in the world we suspect is coming, we can invite them onto a path we are following: to serve honorably, to build relationships with people and the nonhuman world, to develop social, emotional, and practical skills, to be creative and loving on purpose. Older children have keen eyes for hypocrisy. We can admit the gap between our ideals and our practice and model the art of self-acceptance.

Providing children a secure base

Safety is not a question of a lack of threat.
Safety is the presence of connection.

– Gabor Maté[97]

Secure attachment is the best emotional foundation for a child's resilience in hard times, and that requires steady love. Secure attachment is also the best antidote to anxiety for a young child.[98] The best thing we can do for the children in our lives is to show up for them, love them well, and continue to learn and serve as best we know how, while we regulate our own anxiety.

End-stage capitalism has privatized many community functions. This may leave nuclear families alone with little support. Neighbors, grandparents, adopted aunts and uncles can make a big difference by showing up and supporting families. Children with chaotic or lonely lives can sometimes have a transformative relationship with one outside adult. This can be a teacher, relative, mentor, or neighbor who supports and believes in them and can be with them through their personal storms.

Children at different developmental stages require different approaches.

Preschoolers need to know little more than how to explore the world and their bodies through play, to practice emotional intelligence through relationships, and to feel confident that they have a secure base with a parent or other carer (or more than one) who is calm and available. They need healthy relationships, not only with family but with other people and with nonhumans. What you say to them by rote and show them by repetition is what will lodge in their bones. Songs work especially well. Choose your words with care.

Grade school children want to learn and master life skills. They trust you and their teachers to lead them.

- They can hear stories that allow them to imagine taking a constructive place in the world. The default stories of my culture, of superheroes and might making right, of good guys and bad guys, of money and career as the measure of success, will fail them. What stories can we tell them to show them they have a place to live, create, love, and serve in a troubled world? These are essential questions for us to explore.

- We don't have to dwell on troubles, but we can be simple and straightforward when asked about them.

- Children this age can learn hand arts and crafts for living, such as cooking, woodworking, sewing, bike repair, gardening, or other arts. They can have a few skills for a simple and rewarding life.

- Parents of children in high-pressure schools can model respectful limits by setting them around homework—with the school and teachers.[99]

- They can hear our values and see them put into practice at this critical age. That is teaching at its best.

By *high school*, if not before, children in Western industrial societies are pulling away from their parents, doing their best to belong with people their own age. A parent is unlikely to be able to do much educating in a top-down manner. Rather, parents can try cultivating (through example and expectation) a good listener, a skilled negotiator, and a self-directed, thoughtful bearer of some necessary family duties. Punishment and reward can thwart that. Parents must give respect to their teens, and request it in return. They must listen well if they expect to be heard. When struggles happen, they can step away from power plays and instead reveal their feelings and needs, while validating their teen's perspective. The goal, not long off, is a mutually supportive relationship of equal adults.

Consider ways that a teen can experience a rite of passage to capable adulthood. A parent will likely not be able to provide that experience alone.

At 28, my adult child is no longer my child. He is his own person. My time to guide him is done. My job now is to respect him and seek to hear and understand him.

Perspective: Teaching children ecophilia

Peter teaches big ideas to small children in his forest school in the mountains above Palm Springs, California. His favorite big idea is ecophilia: nature has many examples of love and care, and in order to thrive, we must love and care for it back. This isn't a nice abstraction; it's a way of life.

Peter finds that, like most big ideas, ecophilia is far easier to teach to children than to adults. He has observed that when adults enter a forest, they act and feel consciously *other* than the forest. They walk in a stiff and careful way. He describes it as "squeezing through the hole of a doughnut." When children visit a forest, they become the forest, hiding in leaf piles, wading through the brush, collecting found treasures, and wearing a generous coating of forest duff. When they finally emerge, they often bring tales that reveal a deep connection.

Peter can explain ecophilia in scientific terms, listing the interdependent webs of mycelium, plant, insect, and animal, of which we are an integral part. He can also explain ecophilia in human relational terms by recognizing how the nonhuman world sustains us, a form of love and care that calls forth our gratitude and our desire to return that care. We need to know ecophilia not only as a fact but as a living relationship in order to participate in it.

He sometimes hears kids reciting ideas from the wider culture, which he calls ecophobia. They may do this even after they have participated in his forest school for years. "Plants grow far apart," asserted one 13-year-old, "because they compete with each other." This idea was promptly disproved with great hilarity because Peter and the boy were walking through a jumble of plants intertwining with one another.

"It's a dog-eat-dog world," another child said.

"When have you seen dogs eating each other?" Peter challenged.

Peter's advice is simple: *teach your child ecophilia*. The best way to teach ecophilia is by living it. The way to develop ecophilia is to get out in nature. This is easier for some people than others. Yet most of us can manage a windowsill garden and birdwatching at the park, at the very least.

Ecophilia is not the only big idea Peter wants us to know and teach our children. For instance, forests are starkly different from factories; they are emergent and complex rather than planned and controlled. Children's learning, resilience, and creativity are much better fostered in a forest-style school than a factory-style school. Peter is writing a book on education. In the meantime, he has free resources to support parents and teachers at: https://hilltopeducation.com/category/outdoor-learning-activities/

Many voices: What do the old owe the young?

Ever-increasing rates of anxiety and depression among Western youth (from long before COVID) are signs of their despair over the world they are experiencing. While treating individuals is helpful, the causes are systemic. When so many young people suffer, the system they are raised in is broken. Focusing on root causes helps relieve young people of shame or blame. It also invites us to find ways to live that support psycho-spiritual health.

What do the old owe the young in this time of great loss? What words and actions can we offer to support and encourage the teens and young adults in our lives for the hard road ahead? Parents have a special challenge, but all adults can find ways to act as elders to support younger folks. Here are some answers people gave when the question was asked in the Deep Adaptation private Facebook group, *"What do the old owe the young?"*

Apologies:

- Apologies and support for how young people choose to respond or adapt, not nagging lectures.
- Deep, unrelenting apologies.
- I think what I need is a sincere apology, and recognition of the depth of suffering my generation feels now and will feel. I don't need boomers to DO anything other than stop making it worse and take responsibility for failing to act.

Encouraging and supporting:

- Talk less and listen more. Stop instructing and start empowering.
- Encourage them to take the reins.
- Help them to create a narrative and find their purpose and meaning in life.

- Encourage their art, music, and creativity, as I truly believe that the artists will help us emotionally deal with what's coming.

- We owe them opportunities to build the skills that they will need in this uncertain future. We owe them body wisdom practices that will support their own deep adaptation.

Practicing elderhood:

- Learn how to be the elder you wish you had had as a youth, as a young adult, and as a parent.

- Do the psycho-spiritual interior work that will make us worthy of being called elders.

- Younger generations fight the same feeling of guilt [as many of their elders.] We can help them to show the way out by being an example of how to deal with guilt for those things we do not have control over. But first, we have to deal with our own guilt, shame, and grief.

- My college students were racking up enormous debt for the promise of a j.o.b. in the real world. They were working 30 hours a week and taking 18 credit hours. We were training them to hustle for debt repayment and to disregard their well-being. I started asking them what made them feel nourished. They talked about family dinners, friends who care about them, food with an un-complicated backstory, and rest. These are all things they didn't have much access to in college. So we spent time in class creating work that nourished them—solidarity economy projects (figuring out how to buy something cooperatively), ecological projects (designing a food forest), and mutual aid projects (sharing stuff).

- Learn how to be a mentor.

- Teach how to grow food or rebuild soil. Whatever you enjoy about life: teach them that. Share your passions with them.

The word "owe" calls forth ideas of wealth transfer. In the U.K. and the U.S., younger generations have significantly smaller real income and home ownership than Baby Boomers did at their age.[100]

- What do Boomers owe? Money. Yes, money. Reparations. Pick an activist and make them the beneficiary of your life insurance.

- Some Boomers (20% in the U.K.) own multiple properties. More have a nice front yard that they could let local young urbanites use to try out city farming and permaculture. Some have a lot of acreage on their vacation homes that could have a partial use for a tiny home eco-village. Older folks could have a communally maintained grant fund to help young families who want to turn their yards or patios into food forests or pollinator hubs instead of grass. There aren't enough apologies in the world that could replace giving one young person or couple or family a leg up. It's messy, there are legalities, and no one wants to put themselves out there or get taken advantage of...yes, I know. But if not now, when?

- Money, land, and housing would be nice. I know it sounds far-fetched, but seriously.

Being responsible elders

Children have never been very good at listening to their elders, but they have never failed to imitate them.

— James Baldwin

It is time to allow young people to take authority and responsibility for their lives. Our modern ways of life no longer serve them. Our foundational gift to children can be seeking to live as if Earth matters. Additionally, here are some things we can do.

1. Things to avoid.

Avoid saying, "Well, I've had my run. If the world goes to hell, I won't have to live to see it."[101] Likewise, avoid saying, "Your generation will have to step up now and fix things." These things are too cruel. Avoid judging young people. Avoid imagining they are deficient, defective, or incapable. Avoid sounding like you have any fixes or answers.

2. Point out stories that have failed us, and suggest stories that work.

We elders can point out that stories that blame youth for not doing the nearly impossible, or not being able to fix our predicament, are wrong. We can take responsibility for what we were not able or willing to do. The world is beyond saving, but much living, learning, and loving remain. We can offer young people stories in which they can succeed at what is doable, by living those stories ourselves. The most valuable stories affirm the intrinsic worth of every person and the value of creative, service, and caring work.

3. Support and encourage young people to experiment with living and learning from different models than those given to them by our sick culture.

Can we begin by admitting that we are part of, and products of, a sick culture? Can we grant young people the authority to envision cultures and lifeways that may sound absurd to us? Such humility does not come easily, but I believe it is necessary to empower young people.

Meanwhile, we elders can give permission and make opportunities for young people to experience nature, learn creative, household and handyperson skills, and practice ecophilia: developing a relationship with the nonhuman world. Everyone should know about WWOOFing (worldwide opportunities on organic farms), where they can work 4-5 hours a day, for days to months, in exchange for food, lodging, and experience.

4. Lift up the dignity of work and the freedom of living simply.

Working at the shoe store or the grocery store is not a failure. It is honorable work. Making do without things one's parents took for granted is not failure. It shows realism and respect for our Earth home.

5. Give resources to empower young people.

We elders can identify ways to empower youth and give them time or money, as we are able. We can befriend young people, maybe house them for a time, maybe pay some of their expenses to live in ways they feel called to live, or just invite them to stay with us for a while to save money. We can mentor what we know, whether simple living, spiritual practice, social organizing, emotional regulation tools, gardening, craft, or repair...

6. Be compassionate witnesses, deep listeners, and encouragers.

The world of the future truly is in the hands of young people. They may need guidance, but more than that, they need their visions and efforts affirmed, not dismissed. Elders can make sure the dreams, struggles, and stories of the young are heard, honored, and respected. We can trust them, believe in them, and support them as they live into their future.

I've said it, now I get to live it. Fellow elders, what's on your list?

Summary and reflection

Many young adults in industrial consumer society expect (and experience) a world of fewer opportunities than their parents, and looming disasters as well. Elders cannot change this reality, but we can support the children and young people who must face it.

- Ellen wants to be happy, and find simple but important ways to be of service.

- Andrew is prone to grim resignation but works in electoral politics, to help build the world he wants. Staying active feels good, whatever the result.

- As Ari studies watershed management, she is also supporting environmental activists her age and younger. Relationships with other activists sustain her.

- What do we tell the children in hard times? The truth, in gentle and age-appropriate terms.

- Secure attachment, and our own emotional calm, is the best gift we can give children in hard times. After that, what children need varies with age.

- *Ecophilia* is Peter's word for the relationship with nature that Peter teaches in his forest school.

- Many voices speak about what the old owe the young in these times.

- Finally, my reflections on how we who are not young might become effective elders for the young people in our lives.

CHAPTER TWELVE

Planting seeds

Among people learning to respect our Earth home you will find many gardeners. Gardening can be an empowering action and a deep experience of partnership with the nonhuman world. Here are some stories and reflections on gardening as if Earth mattered.

Gardening: Co-creating with the nonhuman world. From someone who hasn't managed it well, an invitation to grow food.

A monk's gift. My garden nurtures seeds from a variety apparently hundreds of years old.

Perspective: Growing food anywhere. Jane can grow food anywhere. So we can too.

Growing a native gardener. My experience creating a garden of native plants.

Perspective: Rebuilding for an uncertain future. Margi has learned hard lessons while rebuilding her home after cataclysmic Australian bush fires.

Homegrown wisdom. Indigenous and traditional knowledge can be transformative if we respect it.

Perspective: Transplanted and growing roots. Brennan is learning to garden to help feed his family and his neighbors.

Gardening: Co-creating with the nonhuman world

Gardening is an active participation in the deepest mysteries of the universe.

Thomas Berry

You don't have to grow all your own food. But from people who have survived hard times, I have learned that growing even a little fresh food can go a long way toward health and a sense of empowerment. Beyond the practical benefits, growing a garden can be a cooperative connection with the nonhuman world that feeds our bodies and souls, and makes us wiser in the ways of the nonhuman world.

So many people would love to teach you how to grow food in ways that replenish rather than deplete the soil. They have systems to teach us that maximize both food yield and our understanding of the rhythms of the nonhuman world. Permaculture and backyard gardening classes are all over the internet; many are free. If you thought you couldn't garden because you don't have a yard, my friend Jane assures me that cramped living with no actual soil is not a barrier to growing food. A few pots of herbs, an avocado pit, or some green onion bottoms, can contribute a spot of green on the kitchen windowsill.

I have not grown much food. What I do grow, the squirrels, rats, raccoons, and rabbits get first. I do manage to grow fresh herbs and lemons, things that don't interest them. And I have a garden of native plants, which is my refuge and joy. For a long time, inattentive as I can be, I believed I was a lousy gardener. But native plants thrive on inattention.

Gardening keeps me humble. I try my best, and then the garden decides what lives and what dies, what thrives and what doesn't. Even experienced gardeners talk about humility. My garden teaches me about dormancy and rebirth, bud and bloom, and going to seed. When something dies, I get to make a new start in an empty patch of soil. Gardening teaches me to pay attention. I am learning to show up more reliably and not to take it personally when the strawberries all go missing. The squirrel clearly enjoyed them.

Don't let fear of failure stop you from gardening. Plants are very forgiving, whether they live or die. Some people love to garden so much they will probably help you if you ask. They need not know anything about collapse. I dared to begin my native garden because of the support of a mentor, my next-door neighbor. Maybe gardening is yet another thing we shouldn't do alone.

A monk's gift

Twenty-five years ago, I planted sweet peas in my backyard. They still grow from volunteer seeds each year, in part shade where few other flowers thrive. The first few sprouts volunteer in winter. Well, I call it winter, but it's hardly cold here. It used to frost once in a while but not anymore. The little shoots lounge about until spring, when they race up the tomato cages I install for their convenience. Soon they are smothering rose bushes and waving interwoven branches at the sky. As legumes, they want no fertilizer, just a lift off the soil and a half-day of sun.

Peak sweet pea season comes in late spring. I harvest hands full of the heavenly-scented two-toned burgundy and purple blooms. I put vases in every corner of the house. I run out of vases and use juice glasses. The sweet fragrance drifts all through the house and yard. I give bouquets to friends, neighbors, and delivery people who visit my house. Some years I get lazy and let them stay on the vine. They then produce less and do not reach their full glory. They are quite clear about their purpose: to bring a few days of divine sight and scent to any who will receive them.

The sweet peas I planted some twenty-five years ago were hybrids with blooms in assorted colors. When volunteer seeds sprouted in following years, they... unhybridized? Reverted to type? Bicolor burgundy and purple blooms came up everywhere, with only a few other colors. I am not the only one to notice this process in sweet peas.[102] The bicolor flowers look just like the Matucana sweet pea, said to have been discovered in 1695 by a Sicilian monk named Cupani. Or in Peru in 1543. Ancient beauty with a will to persist has graced me with annual joy over these many years.

Controlling gardeners cannot abide such things as a tomato cage full of dried pea vines in the long summer days, the pods snapping open and spraying little brown balls like shot across the patio. I hold space for the withered pea vines of summer, knowing their little pea shot is settling in and plotting a return next winter. How fortunate for me that I was not a diligent gardener many years ago, one who snipped every bloom before it could go to seed or ripped out the vines when they began to turn brown. Retro-Matucana returns each year faithfully. He and I have a very agreeable collaboration. I could ask my friend Rachel to run his DNA versus commercial Matucana to learn his factual lineage, but I choose to honor his heritage as I understand it. I hold space for him (space in tomato cages, to be precise) and he puts on an unforgettable show each spring.

Perspective: Growing food anywhere

Grow your food. Or know the person who grows your food.

Vandana Shiva
(when asked what one thing we can do for the Earth)

We've heard from Jane before. She lives in Vermont, and she can't not grow food. That is a powerful invitation for the rest of us who are making excuses for why we can't grow food. She's had a homestead, a lakeside house, a couple of city yards, an apartment, and even a tiny houseboat, and she grew food at all of them.

Jane and her husband Sky were drawn together by their love of growing food, among other things. They created an off-grid, self-reliant homestead on 45 acres in Vermont, complete with sugar maples. For fifteen years, they nurtured it, along with a young family. It got to be a bit much, so they left the country with their school-age children and built a house in the small city of Montpelier, the capital of Vermont. Without intending, they found themselves doing the same thing, though on a smaller scale. They turned their whole yard into a garden. Daughter Dana loved chickens, so city chickens were included.

In 2006 they launched the kids and went to New Orleans to help rebuild homes after Hurricane Katrina. Then they split their time between New Orleans in winters, where they grew vegetables in their yard, and France in summers, where they grew vegetables on their tiny houseboat. Jane had a veggie garden of pots on the bow of the boat, where she grew all her salad greens and herbs, with a tomato plant, a pepper plant, and a cucumber plant.

Extended family issues upended their plans, dragging them back to Vermont in 2010. After a couple of unsettled years, Jane and Sky bought a small plot of land on Lake Champlain and built a home. Their wonderful neighbors offered the use of their land, so they began a small organic vegetable farm.

Jane sees now that Sky was already showing signs of dementia that would take his life in 2021. In just a year he got restless again. He wanted to go back to his New Orleans/French canal boat life. So they sold their new house, but the traveling life didn't work well as Sky's dementia began to manifest more clearly. Finally, they moved back to Burlington, Vermont, sharing a duplex with their son and his partner. Again they gardened, and Sky would volunteer to glean from nearby small farms for the local food shelf.[103]

Their adult children wanted to have a house on a lake, where they and their families could all be together. So Jane bought land on a lake again, and built a simple home, again. Not planning to live there full time, Sky and Jane limited their gardening. On their 3.5 acres, they have twenty fruit and nut trees, plus berries, vegetables, and herbs.

The pandemic changed their plans yet again. They stayed at the remote lake house all year round. Sky could no longer do his carpentry, but he did lovingly tend the young orchard trees until he became too sick. The garden is small by their standards because Jane has been managing it all herself. Only 20 raised beds! Sky died in 2021, and she misses him greatly. She misses the partnership of her planting and his harvesting, but since she cut back to 20 raised beds, she finds it manageable and enjoyable, most of the time.

Jane kept hoping that this current little farm would be that gathering place she and Sky had longed to create, a shelter and a community. She keeps trying to get people to live there. She has room for five or six tiny houses or RVs. But nobody has come yet. The house is isolated, and Vermont winters can be long and cold. Thinking she did not want to be isolated herself, she bought a hundred-year-old house in Burlington about 50 miles away, rehabbed it with help from her adult son, and moved back to that city in the winter of 2022/2023, the third move since 2006. But she loves the quiet and beauty of her lake home and her cats love to roam outdoors, so the house in town is being sold, and she is back at the lake with her little 20-bed garden, growing food.

Growing a native gardener

In 2013 I got hit by a bolt of inspiration. In an uncharacteristic fit of industry, I ripped out my lawn, along with some dull shrubs and half-dead ornamental trees planted by the homeowner's association. I replaced most of my yard with plants native to Baja California and Southern California.[104]

California native plants are different. The plants that grow wild near me don't want soil amendments. Most will happily grow on a mound of decomposed granite.[105] They don't want fertilizer. They never need pesticides. They don't want watering more than once a month, but they do want deep watering in winter. Most of all, they don't want summer water.[106] California natives evolved in a climate where it almost never rains six months of the year, and they have no protection from root rot in warm damp soil. Most of them will tolerate light-to-moderate watering once a month, but some will die from even this much water.

Planting native plants is a different process too.[107] No practical native "lawn" exists; we have to think outside that lawn box[108] to avoid wasteful water use and pesticide and fertilizer runoff that pollutes our bays and oceans. Lush weed-free lawns are one of the largest sources of urban/suburban water overuse and pollution in our area.

California native gardens get no love. I've often heard, "I have a native garden." I look for native plants in that garden and sometimes find a couple. The rest are from deserts or Mediterranean-climate areas around the world, not from California. And no wonder. Ignorance is so rampant it borders on fraud. I once walked into a plant nursery near my home and asked, "Where are the native plants?" The staff member pointed me to a large section of low-water plants. In the whole section, one plant was native to California.

Natives root deep and take a while to express their potential. Many go dormant in the summer, and some look quite dead, so your neighbors will object to your "weeds," and any gardener you try to hire will yank them out. They are so misunderstood! Imagine if we insisted on uprooting every plant that goes dormant in winter. Natives feed birds, insects, and animals. On a minimum of water and no fertilizer or pesticide, their subtle beauty offers a deep connection with the rhythms of nature that are feeding my soul ten years later.

For two years I lived across the country to follow Scott's work. The garden did fine despite getting water only once every six weeks. But I missed my home and my garden terribly. It seems I have been growing roots in that garden too.

For a while now I have been doing weekly Earth Listening meditations in my garden. The plants speak wisdom to me. More precisely, I realize things when I contemplate them. But I like to give them credit. Many times I have heard them relaxing. Industrial consumer society calls that slacking off. They beg to differ. "Leave me alone for now. Let me be dry. Let me lose some leaves and get a bit scruffy. I know I don't look so great now. Sure, go ahead and trim me. Do not expect me to be like the imported and overwatered furniture plants that look the same all year around."

I think of the ever-blooming flower borders at spots in my city, where hired gardeners rip out and replace the plants from one season to the next so that we never have to drive by a plant that is not at peak bloom. Then I think of the young adults I know who go to work after college for big-name tech companies in California for a few years until they, too, are used up, discarded, and replaced with new college grads at the height of their bloom. I prefer the wisdom of the plants native to my area: bloom in your season, and when adversity hits, allow yourself to be dormant, hunker down and grow deep roots.

Perspective: Rebuilding for an uncertain future

Margi and her husband are rebuilding their home and garden after the Black Summer fires in Australia that started in late 2019 and kept raging into 2020. 73,000 square miles of wildfires decimated ecosystems across Australia, and billions of animals died. Half of Margi's island home also burned. For months fires raged nearby until a firestorm finally roared through her property and her community, destroying her home and farm and 88 others in its path. Rebuilding has been slow, with virtually no support from the government. In her book *Fire: A Message from the Edge of Climate Catastrophe*,[109] Margi details her community's heartbreaking experience of barriers to recovering and rebuilding after disaster, largely due to governmental negligence. She is committed to growing food as part of rebuilding. In her words:

> In the wake of this disaster, I am focused on building up our capacity to grow more food than we need, and preparing to survive as long as is feasible. We know the wildfire will come again, and probably inside the next decade. I do this not for myself, but to be here to protect what remains wild and innocent of this travesty—the birds, trees, animals, insects—for as long as I

can. I see my role now as giving their lives space until I can't hold back the tide any longer. I know it's a lost cause, but every life lived is precious and to be treasured.

I know I can't do this alone, so I am investing in close relationships with people who share my worldview. I am fortunate in that my partner and I are 100% in agreement about what's happening, our role in it, and our ability to affect change.

Like so many of our neighbors, we are also rebuilding absolutely everything after a wildfire decimated our world. Am I being pure in sourcing 100% recycled goods in that rebuild? No. But, I did spend 30 years living that life before the wildfire. Now I need a home and infrastructure and I don't have the energy or resources I once did to walk against the tide. The home we've rebuilt is very small, essentially a one-bedroom house. But we have put up one large steel shed, and another will go up in a few months. We need this space to house equipment and for food processing.

We have consciously gone 100% off the grid for power. We were already 100% water self-sufficient. Very little waste leaves our farm. Very few inputs come onto our farm. We are making do and mending what we can.

We no longer fly anywhere. And refuse to be drawn back into that world for work. Although there will be a few domestic flights this year that we will offset as best we can by planting trees across our farm.

Before the fires, we grew and processed a lot of our own food, and already we are back to about 40% food sufficiency. But it takes time to build food production infrastructure. By the end of 2025, we will be growing and processing 95% of it (including grains and legumes)... if we all last that long! The only reason we are not at that stage right now is that the entire garden and everything we had before was destroyed in the fire and it's taken two and half years to rebuild and replant... and nut trees take time to grow.

What food we buy is organic, and usually bulk... but not for virtue's sake. It's what we have always done and we prefer it.

Will any of this make any difference to the outcome? *Absolutely not.* We know that. These are choices we make because they feel right for us. And, it sets us up to be able to do what we can to extend the lives of the wildlife that surround us, for as long as that is possible.

Homegrown wisdom

We never know how our small activities will affect others through the invisible fabric of our connectedness.

– Grace Lee Boggs

When modern people try to use wisdom that has been in the keeping of traditional and Indigenous people, the transfer often does not go well. Ideally, transplanting that kind of wisdom is done with the utmost respect and care, with attention to place, process, and relationship. Uprooting it from its context can transform it from wisdom and healing to folly and harm. A sincere attempt to fully respect a sense of place, process, and relationship makes the whole endeavor so much more helpful. And it can transform the adoption of Native practices into the transformation of people. Isn't that a better goal? Let me give you one example, near to my heart, of unintended consequences and a proposed remedy.

Salvia apiana,[110] White Sage, is the plant of choice used for smudging ceremonies all over the world. The silvery leaves or "smudge sticks" of bundled leaves wrapped in string are burned, dispersing the fragrant smoke as a ritual of purification. What a lovely and healing idea! The range of *Salvia apiana* is limited to fragile hillsides in coastal Southern California and the northwestern corner of Mexico, yet it is now burned all over North America and the world. I am told it is routinely used in schools in New Zealand! So... people come at night in trucks to the places where it grows wild on preserves, public lands, and tribal lands, and uproot whole hillsides of plants, carting them off by the truckload to be sold in New Age stores (and school supply stores?) worldwide. *Salvia apiana* can be sustainably harvested by removing about half of each plant, or propagated and grown as an annual beyond its range. But it is cheaper and easier to uproot and denude a hillside. Just another free-market supply chain calamity among thousands, but a particularly poignant one, considering the intended use of White Sage is for purification.

How to avoid this abuse? You could try to find White Sage that is sustainably harvested. But how do you tell? In any case, White Sage from its small range cannot supply the world. The more common herbs traditionally burned by many Native peoples in North America have similar silvery leaves and are sometimes also called white sage, but are in a different fragrant genus, *Artemisia*. They have much wider distributions and may be less vulnerable to overharvesting. I can tell the difference between *Artemisia* and *Salvia*, but I seldom see *Artemisia* for sale and I doubt if one in a hundred shoppers would know the difference. As long as White Sage is the trend, users participate in denuding hillsides for a ritual of spiritual purification.

What can be done? I grow my own White Sage and it volunteers (grows from seed naturally) in my garden but that's not an option for most people. Over 600 species of sage are found around the world, and several hundred of *Artemisia* as well. What if people found out (through study or experimentation) which smudging herbs grow in their local area or in their garden and used those instead? When a plant is homegrown, the process of growing it matters as much as, or more than, the final product. Rituals do not start when the ceremony starts, nor do they end when it closes. The product cannot be separated from the process, and the process is our lives, lived in connection with our Earth home.

Perspective: Transplanted and growing roots

Brennan lives in rural Massachusetts with his wife and two sons. He is a coach and trainer, mainly working online so that he can spend time with his kids and his garden.

Brennan remembers Earth Day when he was young. Once a year, his school had a recycling push and associated festivities. That show had him believing that "We face challenges, but we've got this." As a child, he lived in a suburban setting with a family that was not outdoorsy. But something in him had always longed for the outdoors. Maybe that came from reading *My Side of the Mountain*,[111] the book his sixth-grade teacher gave him as a gift on his elementary school graduation, and that he has given to his son.

Something lived within him that he hopes lives in everyone: that desire to reconnect with the nonhuman world. He likes to say that we are an expression of "life wanting to life," a phrase he got from Alan Watts.

He received an invitation to read a paper by Jem Bendell. "That was it," he says. "I read the article immediately. I read all 36 pages." He realized, "My children might not live to the age of their natural death. And certainly, they won't live with the level of comforts and security that my ambitions have been out to provide for them."

He started the grieving process. He began to cherish his preschool-age kids, as well as trees and waterways, in a different way. Because his wife's job was taking them to a more rural area, they were able to find a house with a big enough lot to produce food for themselves and their neighbors. They made sure that they spent less on square footage and more on growing space with good sun exposure, good microclimates, good woods, and good water access. They were at their second anniversary in this home when we spoke.

Brennan has found it a privilege to steward this land. His family grows annuals in the front yard. The food forest[112] and the perennials are getting established in the backyard. Brennan sometimes entertains thoughts of building a robust bunker and filling it with all kinds of military-style preserved, dehydrated, packaged Meals Ready to Eat. He muses, "We could survive 'the crash' and we could wait it out, while the people who aren't ready for it don't make it." But then he remembers that this approach wouldn't allow him to recalibrate his relationship with the natural world. It wouldn't allow him to become one with natural systems and rhythms. It doesn't allow him to farm in a way that honors cycles of regeneration. And it doesn't invite him to be interdependent with his neighbors.

Brennan hasn't spent his life learning home building, engineering, or other things his neighbors know much better than he does. So he is clear that his contribution to his local community will be food. As he is able, he will be a food producer for as many of these neighbors as he can. He is learning ways to grow food that are regenerative and durable. He hopes that, whenever it becomes necessary, he and his neighbors can all have a loving shared response to collapse.

As he starts his third season of food growing, he knows he has far to go to master regenerative food production. His neighbors are multigenerational families of longtime residents whose politics are at odds with his. They have not yet welcomed him into their community life. So he continues to deliver his jam jars to them, to plant food, and to honor and love the place he is making a home for his family. He wants to grow roots.

Summary and reflection

- Don't let fear of failure stop you from gardening. Growing even a little fresh food can go a long way toward health and a sense of empowerment. Growing a garden can be a cooperative connection with the nonhuman world that feeds our bodies and souls, and makes us wiser in the ways of the nonhuman world.

- Sweet pea seeds connect me with the past, and know how to survive.

- Jane can grow food anywhere. So we can too. Why not try?

- Sometimes, a garden wants to get grown. My California native plant garden did, and co-opted me into creating it. Now it guides me during Earth Listening sessions.

- Margi knows the future is uncertain after losing everything in the Australian Black Summer of 2020, but she is rebuilding her home and garden so that she can care for her human and non-human neighbors as long as possible.

- Wisdom attends to process as much as product. When adopting traditional wisdom, maybe we are the product. Smudge wisely.

- Brennan had a deep desire to reconnect with the nonhuman world. He and his family moved to a rural area. He is learning to grow food for his family and his neighbors.

CHAPTER THIRTEEN

In the meantime

In this chapter, you will find people and projects investing in an uncertain future. They represent a small fraction of the creative ways people are responding to our predicament. Still, I hope they inspire you.

Creating and being created. This is an invitation to get creative as we expect business as usual to become unworkable.

Libraries are possibilities. People are building libraries of all types to preserve legacies.

Perspective: Guns, anyone? Friends are learning rifle use and safety; not my choice, but I respect theirs.

Things fall apart (and we don't have to). When we accept that falling apart is the order of things, we have a more resilient perspective.

Refuge. Creating refuges for displaced people is a deep need and gift.

Perspective: Restoration is not just for land. After a fire on his land, Jonathan appreciates the refuge his friends and colleagues have provided.

Perspective: In the meantime. In South Florida, Mary Jo is figuring out how to get traction for government action on climate adaptation.

Perspective: How to survive the end of the world. Molly's brilliant list of how to live in uncertain times.

Creating and being created

The only way to make sense out of change is to plunge into it, move with it, and join the dance.

— Alan Watts

Our busy human hands and minds are fully engaged when we are creating. Creating can take us outside ourselves into a sense of timeless flow. Our lives have always been made richer by the creations of artists, crafters, musicians, and dancers, whether amateur or professional. In simpler times, much of what people ate, wore, used, and sheltered in was created, crafted, or processed by their own hands, or by neighbors. This required long-studied skills and clever ideas. When supply chains fray, this style of creating will ease our losses. Creating and learning to create is a wise use of our time and effort in uncertain times.

We may create the seemingly ordinary: dinner, a plan for the week, a note of encouragement or thanks to a friend, or a spring garden. We can begin now to appreciate and value the ability to create these things and take satisfaction in our creations. We may create crafts, both for beauty and enjoyment and for the occupation of our hands and minds, freed from our electronic toys. We may be called upon to create ingenious (or good enough) solutions to the challenges of daily living. We can also create rituals and culture, celebrations and community.

We can give children the space and opportunity to build skills for creation. This is almost entirely lacking in modern schooling.[113] For children and adults, the mastery of one creative skill makes acquiring the next skill a little easier.

As I envision living without fossil fuel or long supply chains, or "living at the tideline" in whatever external circumstance, I begin to realize all the creativity and knowledge that is required. Without the massive machine of modernity, its mass production, and sufficient money as its lubricant, people in each local community must create much of their sustenance and culture. Traditional people have managed to do this in thousands of different ways, depending on what was available and needed in different local environments. The stakes were high. My one weekend at a traditional skills gathering helped me gain great respect for some of those skills. I am far from mastery in most craft skills but put me in the kitchen or the garden, and I can create something. Well, co-create. Earth does her part.

Libraries are possibilities

Without libraries, what have we? We have no past and no future.

– Ray Bradbury

Libraries are resources for survival. Sharing and learning what we know and do is the basis of all human cultures.

The innate desire and ability to learn from one another allow us to build up libraries of knowledge and technology. This ability, not our individual intelligence, is the one real superpower humans have that other creatures do not,[114] that has allowed us to span the globe and yes, strain it to breaking. Parts of the knowledge we have accumulated may be best lost, but some knowledge will be crucial for people making their way in an age of chaos and decline, and whatever comes after. So what would you like to learn? You may know of someone who is a whiz at upscaling old clothes or cooking with beans. If not, there's always YouTube, at least for now. Or maybe it's time you took up an instrument, to amuse yourself when your streaming music service fails. Or take a course on something you have always wanted to know at the online khanacademy.org, a resource that is empowering young people all over the world to learn more than their schools can teach. You could invite a friend or an elder to teach you a skill. Then you'll be building a relationship at the same time.

One long night some dozen years ago, I imagined what would be lost when the seas rise to their full extent. This was before climate weirding when I still imagined life would be somewhat normal *before* sea level rise. All the icecaps melting means a sea level rise of roughly 215 feet, 66 meters. So much would be gone: farmland, manufacturing centers, cities, and whole states would be gone. What chaos would be part of that upheaval? I choose not to dwell on this. But what would survive? What is required for a technological age? How long could we cobble together computers from spare parts? What is required to preserve medicine, literature, or music? Would our society become like we imagine the so-called dark ages of Western Europe, with the writings and technologies of the previous age lost or in remote libraries inaccessible to all but a learned few?

Since that sobering night, I have learned of other kinds of libraries. Seed libraries help us grow food under extreme conditions.[115] Skills libraries can exist among neighbors, where you can request a bicycle tune-up or learn about edible wild plants in your area. Neighbors and public libraries sometimes create tool libraries as well, so many people can share that tool they each use once or twice a year.

The myriad nonhuman entities around us are another library from which we can learn. The wisdom literature of the ancient Greeks, the Hebrew Bible, and many ancient and traditional cultures offer stories and aphorisms in which nonhuman protagonists are the wise ones from whom we learn. The physical and conceptual separation of people from the nonhuman world has caused us to ignore this staggeringly diverse and intricate wisdom library. Only in the last years of the twentieth century did ecological studies invite modern people to learn again from the nonhuman world.

Creating libraries and archives of all kinds of human knowledge is a wonderful way to serve humanity in a time of loss. It is also a serious challenge when so much of our world is digital, and digital information is very fragile and decaying as we speak.[116] What things are on your short list for your own library, things that you can offer to family, friends, and neighbors? They might be life-saving skills, but more likely, they will be little somethings to make life more fun or comfortable when our computerized, fossil fuel-powered lifestyle becomes shaky. How could you organize and share your contributions? You don't have to get it right the first time. You could just start building a little library.

Perspective: Guns, anyone?

Two friends in Southern California would like their left-leaning fellows to consider community self-defense.

One friend got serious about self-defense not long ago as he watched hate crimes multiply. He was a target of harassment based on his ethnicity. And he is pretty convinced that the road to collapse will include civil unrest or fascism.

He believes self-defense is a holistic process that includes awareness and knowledge, and an acknowledgment that people can act in awful ways. The best community self-defense is prevention. But preparing to avert threats, he believes, is only sensible in the U.S., where so many people have guns. He thinks some level-headed people in the community should own guns and learn how to use them, including all the safety measures, for defense. It's not about him liking guns. He hates U.S. gun culture. He just wants to prepare for collapse. If not guns, he thinks we all should at least have and know how to use pepper spray or other nonlethal approaches to self-defense.

These friends recently trained with a gun at a range in the high desert about a hundred miles from home as part of a leftist firearm organization. My friend reported the sound waves reverberated in his body, triggering an adrenaline response. Firing a gun was an experience of overwhelming power. He notes that he was taught a very rigid sequence of gun safety rules. One is obvious: never point a gun at a person, even if you're sure it's empty. He does not want anyone and everyone to have a gun, only those who are cautious and responsible.

These friends want at least a few people with guns to share their values. Those people probably won't include me, but they're ready to help me if I request help.

Things fall apart (and we don't have to)

Hope is not the conviction that something will turn out well, but the certainty that something is worth doing no matter how it turns out.

— Vaclav Havel

Nobody ever taught me that things fall apart. I somehow imagined the institutions I valued would keep going indefinitely. I remember the eerie feeling I had years ago when I read the classic book, *Collapse,* by Jared Diamond.[117] He described cultures around the globe that had collapsed and why they might have collapsed. But I didn't think I would see such things in my lifetime.

After contemplating our predicament, I now expect that things do fall apart. It is the order of things. This is a great comfort when I find myself witnessing things falling apart. Instead of becoming disoriented and panicky, I say to myself, "Right. It looks like something's collapsing. Wow, that's sad. No cause for panic." My world holds steady instead of upending. *I* don't fall apart. Usually. Or not entirely, and not for long.

The regional head of a large volunteer organization I dearly love and value, one that I have invested in deeply, admitted the state of things to me in private. All over the country, they can no longer field teams of volunteers to do the core work of the organization. I thought, "This is a collapse in action." Nobody has failed. Nobody's to blame. That organization is just collapsing. I had a good cry, and I'll probably have more. I will plan accordingly. I will not exhaust myself trying to raise the dead or raging against things dying.

Recently a friend flew across the continent, something she doesn't like doing, to support her father after he had a stroke. It was a mess. He refused her help, though he needed it badly. After she gave up and came home, she and I talked about how things collapse and people collapse too. It's in the order of things. She had noticed that a personal collapse like her father experienced had some similarities to the social and ecological collapses she has been watching unfold. She was amazed to realize that she had felt mostly calm and relaxed throughout the dreadful visit. She had brought a whole arsenal of medications to prepare for all kinds of bodily stress reactions that she had suffered at times in the past. She needed none of them. She wasn't trying to fight reality, nor was she expecting it to be easy or to get her way. She could watch things fall apart for her father and accept that she couldn't hold them together. She didn't have to fall apart too.

Refuge

Housing is one of the resources we need that is threatened by collapses. California homes lost to fires make the news worldwide; over 11,000 homes were lost in the Camp Fire in 2018. A year later, only 11 had been rebuilt.[118]

California has a population of about 40 million. Loss from fire is just the dramatic tip of the iceberg of a growing housing crisis here. Housing affordability and supply have been decreasing for years. Tent cities under highway bridges and in parks are the visible signs, but tens, perhaps hundreds, of thousands of people living in their cars or in garages, on sofas or living room floors, are the invisible unhoused. Some leave the state, but you have to have money (or a bus ticket from a sheriff who uses bus tickets to reduce homelessness in his jurisdiction) to leave and take your statistic elsewhere.

This is only the beginning. As disasters and economic loss displace more and more people, places to shelter will be needed more and more. In the U.S., insurance from disaster is already a cruel joke,[119] if it is available at all.[120]

Where could you take refuge if your current home was destroyed? Or if you could no longer afford to keep it?

Now turn that around. Who would you shelter in your living space, and under what circumstances? For how long?

These are personal questions that challenge us deeply. As I contemplate needing refuge myself, I find myself more willing to provide refuge in my own home.

All of us can have conversations with friends and family, far and near, about possible refuge after having to leave home for any reason. Given the wildfires near me in the past few years, I don't need to share my views on collapse to have that conversation. Those of us with housing to spare can take simple steps to prepare to host. We can free up space and set up existing space for comfortable co-housing. We can experiment by offering a spare bedroom to a college student or visitor for a few weeks or months.

Anyone can offer a stay of a few days. A longer stay reveals which spaces and resources are at a premium. Kitchen and refrigerator space were at a premium in our case. The host/owner can be at a considerable power advantage over the guest/renter in things like the use of common areas and storage space. I suggest beginning with an expectation that conversations about arrangements will happen regularly. Boundaries can be set and negotiated. Expect and acknowledge challenging situations and be willing to experiment with novel solutions. And look for ways to enjoy each other's company too. In many traditional cultures, welcome and shelter have been sacred gifts, and sacred responsibilities. We may relearn this truth.

Perspective: Restoration is not just for land

Jonathan has studied eco-psychology, but he made a career practicing more mainstream psychology. Finally, in 2019, he was ready to follow his earlier vision. He put everything he had into a project of beginning an ecosystem regeneration camp, which also offers psychological healing for people through connection with the land. Then the fires came. He found refuge with kindred spirits who had very simple shelters to offer.

The place Jonathan chose to do ecosystem restoration is on the slopes of Mount Shasta in northern California at Hotlum. While doing his pre-purchase research, he learned from the wise people at Calfire (the California Department of Forestry and Fire Prevention) that it was damaged due to logging and mismanagement and prone to severe fires fueled by the overgrowth of brush. This is part of a larger decimation of ecosystems and lifeways that destroyed the habitat of beavers, salmon, deer, elk, bears, and wolves, along with rivers, forests, and the livelihoods of Indigenous peoples.

Since moving there, Jonathan has been evacuated because of fires five times in three years. One of those fires, the Lava Fire, burned the forest he had come to love and planned to restore. In the firestorm, almost every single tree burned. Months later, record rainfall caused washed-out roads and severe land erosion. When I interviewed him, Jonathan had been working to recover for a year and a half.

In the early days of loss and disorientation, he sought refuge with people who were doing ecosystem regenerations like his. He was friends with the director and board members of the Camp Fire Restoration Project near Paradise, California, where almost 19,000 buildings were destroyed and 85 lives were lost in 2018. They said, "Come stay with us." So he did. Next, he stayed with some friends who operate Eco Camp Coyote on the outskirts of San Jose, California. Then he was invited to stay with some friends who run an ecological restoration project along the railroad tracks in the small town of Weed, close to his property that burned.

All these invitations came from kindred spirits in communities where he was already networked. They had been encouraging each other in ecosystem restoration work. This work includes living lightly on the land and cultivating generosity as part of nurturing a natural system. All of Jonathan's hosts share some personal resonance with his experience. Some are tending land that has burn scars. Some have watched their land burn. And most of them recognize their land is prone to burning. They are realistic about the risks, and they can feel despair at times, but that doesn't stop them, he says. "We just get up and brush ourselves off, and we do the next right thing. We don't have to hope that we're going to live happily ever after to have fortitude. I think the right thing is to contribute to life in all forms. And that takes fortitude."

Those short stays with kindred spirits deeply supported him. He felt like he was welcomed into friendly little refuges. The homes were very spare: a yurt, trailers, and other makeshift shelters. The welcomes were personal and informal. The shelter was both physical and emotional. As I write, he is staying at a friend's home at no charge. "That's community," he told me.

Perspective: In the meantime

Mary Jo lives in West Palm Beach, Florida on "high ground," twelve feet above sea level. At age 84 when I interviewed her, she doesn't have the energy she used to, but she does have passion. She is still starting conversations and bringing people together to build community and resilience.

Mary Jo considers herself an accidental environmentalist. While wrapping up another career, she studied contemporary art that addresses social issues and social change. She arrived in Florida with her new Contemporary Art History degree in 2004, two days after a hurricane. She drove through streets filled with debris and darkened stoplights to a new home with broken windows. Another hurricane followed three days later, ripping through the broken windows and destroying the floors of her home. That was the first time that West Palm Beach had been hit by hurricanes in 65 years. It was her wake-up call. Soon after, she learned that Lake Okeechobee was a sewer and that the coral reefs off Florida's coast were dead or dying.

In response, she started a nonprofit called EcoArt South Florida and ran it from 2007 to 2015. Eco Art incorporates ecological principles (and often incorporates ecosystems) as part of art. These imaginative artists challenge our thinking about what is art and what is ecological management, or civic engagement.[121] Projects include public art installations that generate electricity, park landscapes that filter runoff water and provide habitat for birds, and performance art in which the artists help people coin words to describe ecological distress. Mary Jo began by bringing an accomplished ecological artist to South Florida to work with and mentor local artists for several months, finishing with a showing of all their works. She then started meeting with local municipalities and counties across Florida to champion ecological art as public art. Instead of installing a statue in a park, she urged them to commission a work of art that creates habitat, cleans water, or creates energy. She realizes these installations are small and temporary fixes, and do not address our grim fate. But they can help us and the wild places and creatures they may shelter, in the meantime.

When Mary Jo attempted to retire, she started doing research for a book about Eco Art. Recently she accepted that we will go through various collapses, probably ending with one grand collapse. In the meantime, she wonders, how can she be of service? She has set the book aside for now and ponders more local and practical projects. She sees people talking about individual resilience and survival, but she wonders what could be done for wider communities. She knows that trying to redirect government policy is slow and frustrating. Yet her regional and local governments have an impressive emergency management function with paid staff and significant money to spend. Emergency management is already a priority in Florida. People care about their safety and the safety of their loved ones, and most Floridians have been through hurricanes. She believes emergency management is a good place to start conversations about adaptation in the face of climate chaos.

It is taking a while for her to find her way in this new project. Our accountability group that meets on Zoom every other week has been helpful. She enjoys researching to discover what resources are available. Using connections with government agencies and environmentalists that she made with EcoArt South Florida, she is starting conversations that hopefully help people think outside the box. She thinks building local community is a great place to start for resilience.

Mary Jo is still figuring out her role in this new arena when she thought she'd be retired. She has been a producer all her life, making things happen, starting in her childhood with backyard circuses. She hopes at this stage not to be in charge, but to plant seeds and make connections, to help make things happen that will help people survive the next collapse or two in her very vulnerable community and state.

Perspective: How to survive the end of the world

Unitarian Universalist minister Molly Housh Gordon lives in the U.S. state of Missouri. In her 2019 Climate Strike sermon, "How to Survive the End of the World," she wrote this "totally serious list of how to survive the end of the world."[122]

> Get to know your neighbors. Feed them. Let them feed you. Watch each other's kids, grandkids, pets.
>
> Develop the muscle of generosity like you are training for a giving ultra-marathon. Share everything you can with anyone who asks, and ask for what you need.
>
> Get in touch with your body. You will need it, and it knows things. Pay attention to what is happening below your neck.
>
> Tell the truth. Tell it to yourself first.
>
> Sit at the feet of your most vulnerable neighbors and your own most vulnerable places. They have the most to teach you about survival. Listen.

Remember your ancestors, and the things they survived. Find the resilience that is your birthright and the courage that made way for your life.

Practice taking risks. Show up in every struggle where someone is fighting for their dignity, because that is how we will all survive.

Learn about reparations and native sovereignty. Double down on exorcising supremacy systems from your soul.

Learn to be tender. Refuse to be hardened. Let your heart be moved. Every damn time.

Root in the place you are. Learn its history. Learn its geography. Learn its seasons.

Sing. A lot. And dance. Make art. Make love. Rest luxuriously. Eat pie.

The world is ending and beginning now. We are surviving now. Let us love, let us connect, let us fight like hell for the dignity of each and all.

Summary and reflection

- When times get tough, creativity will allow us to make life more tolerable. In the meantime, it's fun. Let's skill up in crafting and creating.

- All kinds of libraries exist, for knowledge, for seeds, and more. They contain gifts from the past and gifts for the future. Creating a library is a worthwhile task for an uncertain future.

- A friend is learning rifle use and safety with his socialist friends; not my choice, but I respect his.

- Knowing that things fall apart, we need not panic when they do. When we can be calm, we are fully present to best respond to the challenges we face, and we can support others in the storm.

- Prepare to give and receive refuge, a place to live for a while, for when we need it.

- Jonathan wanted to do ecosystem restoration, then he needed restoration, and his friends doing restoration work came through for him.

- At 84 years old, Mary Jo is making connections, seeking to help emergency management directors in South Florida become relevant and effective.

- Finally, Molly offers generous ideas about how to find meaning and joy at the end of the world.

CHAPTER FOURTEEN

Endings are beginnings

In a time of destruction,
create something: a poem, a parade, a community,
a school, a vow, a moral principle;
one peaceful moment.

– Maxine Hong Kingston[123]

As this book ends, how will you continue to cultivate your calm and compassionate center? Being the eye of the storm might mean living like this:

- Admitting the reality of the complex intertwined crises undermining our Earth home, and navigating the complex feelings that arise from that knowledge.

- Accepting uncertainty about what the future will bring.

- Letting go of stories that invite further destruction; instead lifting up stories that don't require happy endings, but offer beauty, meaning, and purpose in the face of loss.

- Anticipating, bearing witness to, and grieving loss and pain, human and nonhuman.

- Finding meaning and purpose despite losses, without resorting to denial or quick fixes.

- Cultivating and sharing a non-anxious presence that will help ease the fears of others, using a toolbox of emotional and spiritual practices.

- Staying connected with people: neighbors who support each other in living, and kindred spirits who support each other in living out our values.

- Looking ineffectual or foolish by the standards of industrial consumer society, because you live by different stories than those upholding industrial consumer society.

- Respecting limits, including the limits of our own bodies and minds.

- Savoring and nurturing life and love, beauty and wonder, courage, and compassion, no matter what happens.

- Experimenting with ways of living simply, ways that respect humans and nonhumans near and far, ways that you enjoy.

- Taking action to serve and love humans and nonhumans, generously and authentically, without expecting to save the world.

Your way of being the eye of the storm in these strange times will be unique. There is no one right way to be or act in the mapless territory we have entered. I hope that the lists, ideas, and conversations in this book inspire you. Remember that you need only take in a few at a time. I trust that you can find portions to support you now, and set aside the rest of this book to consult later.

You may want to reflect on whether your stories about industrial consumer society and your place in it have shifted while reading this book. Perhaps you question a story that seemed to you to be fact. Or you now have affirmation of a hunch that you had never heard voiced before. You may find that your worldview continues to shift over time.

Remember to find your people, and don't give up if it takes a few false starts. If you are still alone in your awareness of collapse, one approach to finding kindred spirits is to start a reading group of this book, online or in person, with people who care about ecological and social justice.

Throughout the book, people have noted that action can be calming. I invite you to take small actions to break out of any routine that is not serving you. Walks, conversations, with neighbors and possible kindred spirits, communing with the non-human world, cooking a new recipe, meditation, or journaling are some ideas. Notice I did not include social media on this list.

Please take your time to experiment before committing to a life change or a large project. Drastic action before deep reflection is not the goal; it's what got us into this predicament. Think and explore long before starting any large project from scratch. It's much more fun, and less work, to become a supporter of an existing project. Don't expect to make a living with your efforts; instead, make a life. Remember that tending, sustaining, and befriending are all honorable work. And remember that tending the back end, whether composting your waste or doing administration of groups, is also love in action.

—〰—

Anxiety is contagious. Calm is contagious. And courage is contagious. Our lives ahead will almost certainly be deeply challenging as we face heartbreaking and disorienting losses. Yet, in the company of others, we can imagine and start living into a future that respects limits and treats humans and the nonhuman world with dignity, whatever unfolds. We can create that world, at least in fragments. With the tools and references in this book and with kindred spirits, we can create calm, purpose, gratitude, and even joy in our lives now, and we can share these gifts. We can be the eye of the storm, creating calm and courage to shelter others.

—〰—

If you have found this book helpful, please share it. And remember that endings are also beginnings.

Epilogue

Leaving an oasis of pine and cool breeze,
We descend into dry canyons dotted with scrub.
We are provisioned with abundant gifts of
Beauty and wisdom, birdsong and soft woodland duff,
Perspective and resonance, cold water and ancient fossils.

We are full.
We are affirmed as co-creators breaking out of imposed chains,
Alive to the presence of tiny treasures at our feet,
Deep canyons of meaning that have no words,
Senses that grant superpowers if we but attend.

The canyon is in us.
The birdsong is in us.
The charred sentinel pines are in us.
The seeping water and the squishy mud are in us.
All is beauty
All around us
And in us,
All Earth's precious jewels.

ANNOTATED BIBLIOGRAPHY

I have not read all of the many books published on topics related to this book, so this bibliography cannot be complete. Some web references in the text are in the notes but not listed here. Many of the older books listed here can be borrowed as PDF versions at the internet archive: archive.org.

Bacigalupi, Paolo, "A Full Life." *MIT Technology Review*, April 24, 2019, https://www.technologyreview.com/s/613349/a-full-life/. This is a short story about climate collapse with an important moral: consider the children.

Bendell, Jem, "Deep Adaptation: A Map for Navigating Climate Tragedy." Insight Home, University of Cumbria, July 27, 2018. Revised 2nd Edition released July 27, 2020, https://insight.cumbria.ac.uk/id/eprint/4166/. Downloaded by over a million people and translated into several languages, this article was influential in helping people from many walks of life admit and name the reality of our predicament. For more information about this article and responses to it, see the Wikipedia article: Deep Adaptation.

Bendell, Jem and Read, Rupert, *Deep Adaptation: Navigating the Realities of Climate Chaos.* Cambridge, UK and Bedford, MA, USA: Polity Press, 2021. This academic book of essays gave legitimacy to the topic of deep adaptation and contains the 2020 version of the title essay.

Bendell, Jem, *Breaking Together: A Freedom-Loving Response to Collapse*, Goodworks, 2023. An examination of a variety of dimensions along which societal collapses are unfolding, with the perspective that freedoms are essential to reducing harm.

Bendell, Jem and Katie Carr, Katie, "Group Facilitation on Societal Disruption and Collapse: Insights from Deep Adaptation," *Sustainability* 2021, *13*(11), 6280. https://doi.org/10.3390/

su13116280. This article describes (in academic language) the benefits of group facilitation on topics related to our predicament.

De Causmaecker, Sven, Svenergy, https://svenergy.info.

Dwinell, Jane, *Freedom Through Frugality: Spend less, have more.* Addison, VT, USA: Spirit of Life Publishing, 2010.

Fleming, David, *Lean Logic: A Dictionary for the Future and How to Survive It,* Ed. Shaun Chamberlin. https://leanlogic.online or you can get the gorgeous coffee table book, White River Junction, VT, USA: Chelsea Green, 2016. It is more like an encyclopedia, full of challenging and delightful, practical and philosophical wisdom on how our descendants might find a way of living in harmony with the Earth and each other.

Fox, Matthew, *Original Blessing: A Primer in Creation Spirituality Presented in Four Paths, Twenty-Six Themes, and Two Questions.* Santa Fe, NM, USA: Bear & Company, 1983, and more recent editions. Scholarly and inspirational essays that recover and redeem the history of Earth-centered Christianity, and propose a framework for contemporary spirituality that honors the Earth.

Friedman, Edwin H., *Generation to Generation: Family Process in Church and Synagogue.* New York: Guilford Press, 1985. This is a classic text for dealing with groups under stress. The issue is not the issue. The issue is basic human needs and relational dynamics.

Ghosh, Amitav, *The Nutmeg's Curse: Parables for a Planet in Crisis.* Chicago, IL, USA: University of Chicago Press, 2022. The title parable is a powerful illustration of the brutality of Western colonial commerce and the inextricable linking of human oppression with environmental destruction.

Graeber, David and Wengrow, David, *The Dawn of Everything: A New History of Humanity.* New York, USA: Farrar, Straus and Giroux, 2021. The authors use anthropological and archeological records in addition to more typical historical sources to build a powerful case for the diversity of complex cultures that have existed around the world, and to make the case that some complex societies have not been based on hierarchy or control as industrial consumer society is.

Hagens, NJ, and White, DJ, *Reality Blind: Integrating the Systems Science Underpinning Our Collective Futures,* Volume 1. Independently published, 2021. Available in bookstores or at: https://read.realityblind.world/view/975731937/i/. This is the text for "Reality 101," taught by Hagen at the University of Minnesota. Energy is the foundation of our civilization and of life itself. We are burning millions of years' worth in a few short decades. This is a thick book, but the authors work through many important technical concepts related to fossil fuel use with user-friendly language. A good alternative to Murphy.

Henrich, Joseph, *The Secret of Our Success: How Culture Is Driving Human Evolution, Domesticating Our Species, and Making Us Smarter.* Princeton, NJ, USA: Princeton University Press, 2015. Humans are unique among animals because we are compelled to copy one another. That mimicry is the powerful foundation of culture. And culture is extremely malleable.

Hine, Dougald, *At Work in the Ruins: Finding Our Place in the Time of Science, Climate Change, Pandemics, and All the Other Emergencies.* White River Junction, VT, USA: Chelsea Green, 2023. Hine's artful and insightful writing tackles the role and limitations of science (and of politics) in navigating our predicament. The topic sounds dry, but the writing makes it sing for me.

Jacobs, A. J., *Thanks a Thousand: a Gratitude Journey.* New York: Simon and Schuster/TED, 2018. An entertaining and deep exploration of long supply chains and interdependence, all from a cup of coffee.

Jenkinson, Stephen, *Die Wise: A Manifesto for Sanity and Soul.* Berkeley, CA, USA: North Atlantic Books, 2015. An invitation to step beyond the absurd denial of death in Western culture, from a practical philosopher and storyteller. Many have found this book helpful.

Johnson, Trebbe, *Radical Joy for Hard Times: Finding Meaning and Making Beauty in Earth's Broken Places.* Berkeley, CA, USA: North Atlantic Books, 2018. A call, stories, and detailed, practical instructions for engaging with the broken places of our Earth home.

Johnson, Trebbe, *Fierce Consciousness: Surviving the Sorrows of Earth and Self.* Ithaca, NY: Calliope Books, 2023. In short essays, this book illustrates being the eye of the storm. Radiant.

Karen, Robert *Becoming Attached: First Relationships and How They Shape Our Capacity to Love.* Oxford, UK: Oxford University Press, 1998. This a well-researched and easy-to-read invitation into the history of *attachment theory*, the idea that children bond to their caregivers, and that bond is crucial to a child's emotional and mental development. Perhaps as interesting as the (very interesting) research documented is the fact that Western scholars needed to do research to believe that attachment theory existed. Talk about ivory towers!

Kimmerer, Robin Wall, *Braiding Sweetgrass: Indigenous Wisdom, Scientific Knowledge and the Teachings of Plants.* Minneapolis, MN, USA: Milkweed Editions, 2013. Essays and stories that offer a window into the science and sociology behind indigenous lifeways. A favorite of many I meet.

LePage, Terry, CA Native Garden, http://canativegarden.blogspot.-com. A California native garden blog, with lots of pretty pictures, and instructions on what and how to plant, and how to care for coastal California native plants.

Longacre, Doris. *Living More with Less*, 30th Anniversary Edition. Scottsdale, PA, USA: Herald Press, 2010. Many books have been written on living more with less. This one is simple and has stood the test of time.

Machado de Oliveira, Vanessa, *Hospicing Modernity: Facing Humanity's Wrongs and the Implications for Social Activism.* Berkeley, CA, USA: North Atlantic Books, 2021. The author builds careful arguments and uses stories of colonialism, racism, and indigeneity to challenge the axioms of modernity, and to remind us that many traditional people have survived or are surviving collapse.

Macy, Joanna and Young Brown, Molly, *Coming Back to Life: The Updated Guide to the Work That Reconnects*, revised edition. Gabriola, BC, Canada: New Society Publishers, 2014. A manual for The Work That Reconnects, created by Joanna Macy and collaborators. A conceptual and practical guide to workshop/retreat style engagement with grief and re-storying. See also the website: https://workthatreconnects.org/resources/about/.

Martin, Laura, *Breaking into Light*, Washington D.C., USA: Politics and Prose, 2023. Poems for spiritual sustenance, from a Christian faith that emphasizes social justice.

McNaughton, Elizabeth and Wills, Jolie, *Cards for Calamity: Lift the lid to navigate life after disaster.* Denver, CO, USA and Napier, New Zealand: Hummingly, 2019. These cards are an invaluable resource for restoring and maintaining the most important thing besides basics like food, water, and shelter: a clear and calm mind.

Menakem, Resmaa, *My Grandmother's Hands: Racialized Trauma and the Pathway to Mending Our Hearts and Bodies.* Las Vegas, NV, USA: Central Recovery Press, 2017. This book looks at the effects of racism on bodies, White, Black, and blue (law enforcement.) It offers a vocabulary of embodiment and many practical (embodied) tools for facing and processing difficult experiences. For many people, it has been transformative.

Moss III, Otis, *Dancing in the Darkness: Spiritual Lessons for Thriving in Turbulent Times.* New York, USA: Simon & Schuster, 2023. This recent slim book of essays illustrates the life-giving spirituality of Black Christian Americans in the face of violence and oppression.

Murphy, Thomas W., Jr., *Energy and Human Ambitions on a Finite Planet.* UC San Diego Open Educations resources website, 2021. https://escholarship.org/uc/item/9js5291m, https://doi.org/10.21221/S2978-0-578-86717-5. This free and comprehensive physical science textbook should be required reading for anyone working in climate or sustainability. If it were, much ignorance, absurdity, and obfuscation would be avoided. Murphy teaches at the University of California at San Diego.

Orlov, Dmitry, *The Five Stages of Collapse: Survivors Toolkit*. Gabriola Island, BC, Canada: New Society Publishers, 2013. With wry humor, Orlov examines the USSR's 1990s collapse as a possible guide for other societal collapses.

Prideaux, Margi, *Fire: A Message From the Edge of Climate Catastrophe*. Mile End, Australia: Wakefield Press, 2022. A factual and deeply personal account of the burning of Kangaroo Island during Australia's Black Summer of 2019/2020, when vast tracts of that nation burned. Not an easy read, but full of practical information for those who wish to restore devastated communities or understand the political and bureaucratic barriers to doing so, or better understand what natural disasters exacerbated by climate chaos can do.

Roberts, Elizabeth and Amidon, Elias, *Earth Prayers: 365 Prayers, Poems, and Invocations from Around the World,* also published as *Earth Prayers from Around the World.* New York, USA: HarperOne, 1991. A classic, used by many people to cultivate reverence for our Earth home.

Robin, Vicki, and Dominguez, Joe, *Your Money or Your Life: 9 Steps to Transforming Your Relationship with Money and Achieving Financial Independence.* New York, USA: Penguin Books, 1992. A classic program that helps a person budget, plan, invest, and make choices about money that use their life energy in ways that are meaningful to them, instead of as a slave to money, work, and consumption.

Rosenberg, Marshall, *Nonviolent Communication: A Language of Life,* 3rd Edition. Encinitas, CA, USA: PuddleDancer Press, 2015. The founding text of Marshall Rosenberg's Nonviolent Communication. Lots of examples are given in his unique style. Learning to speak using Nonviolent Communication requires more than this book, but it is a great foundation.

Servigne, Pablo and Stevens, Raphaël, *How Everything Can Collapse: A Manual for our Times*. Cambridge, UK and Bedford, MA, USA: Polity Press, 2020. First published in French, 2015. This book presents facts, extrapolations, and the likely social and emotional impact of the collapse of industrial consumer society in accessible language, with clarity and a gentle tone. The authors have helped created the field of collapsology. Written in 2015, it underestimated the rapidity and devastation of unfolding climate-related disasters but addresses other predicaments with clarity.

Thurman, Howard, *Jesus and the Disinherited*. First published in 1949. Boston, MA, USA: Beacon Press, 1976. A perennial jewel that presents an essential message of the Christian Gospels and explains how they have given life and dignity to oppressed Black Americans.

Trungpa, Chogyam, *Cutting Through Spiritual Materialism*. Boston, MA, USA: Shambala Publications, 1973, and more recent editions. This book kept me Christian by helping me do exactly as the title states.

Weller, Francis, *The Wild Edge of Sorrow: Rituals of Renewal and the Sacred Work of Grief*. Berkeley, CA, USA: North Atlantic Press, 2015. Many people who are learning to deal with grief at the state of the world cite this book as a game changer.

Wray, Britt, *Generation Dread: Finding Purpose in an Age of Climate Crisis*. Toronto, Canada: Knopf Canada, 2022. Wray uses a journalistic format to examine questions close to the hearts of young people in North America with rigor and compassion. She also reveals her personal soul-searching about whether to have children.

Zimmerman, Jack and Coyle, Virginia, *The Way of Council*, 2nd edition. Spring City, PA, USA: Bramble Books, 2009. A story-based guidebook for using the Way of Council as taught and advocated by the Ojai Foundation.

NOTES

1. Several early readers asked, "But what if the reader is to blame?" Here is my reply. In one sense, anyone who has benefited from industrial consumer society is to blame. But since we are trapped in a sick system, what good is blaming anyone? Why not inspire them to try something different instead?

2. This resource may be of help if you have been through a disaster: McNaughton, Elizabeth, Wills, Jolie, *Cards for Calamity: Lift the lid to navigate life after disaster* (Denver, CO, USA and Napier, New Zealand: Hummingly, 2019).

3. Peggy McIntosh, "White Privilege: Unpacking the Invisible Knapsack," originally published 1989, SEED: the National Seed Project. https://nationalseedproject.org/Key-SEED-Texts/white-privilege-unpacking-the-invisible-knapsack.

4. The Deep Adaptation Forum (DAF) offers events (mostly free) and online resources for people who are seeking or building supportive communities to face climate and social devastation. You can learn more at deepadaptation.info. **Deep Adaptation** is a concept, agenda, and international social movement. It presumes that extreme weather events and other effects of climate change will increasingly disrupt food, water, shelter, power, and social and governmental systems. These disruptions would likely or inevitably cause uneven societal collapse in the next few decades. The word "deep" indicates that strong measures are required to adapt to an unraveling of western industrial lifestyles. The agenda includes values of nonviolence, compassion, curiosity and respect, with a framework for constructive action. For a fuller description, see https://en.wikipedia.org/wiki/Deep_Adaptation.

5. Carol Bialock, "Breathing Under Water." *Coral Castles* (Oregon: Fernwood Press, 2019). Used by permission.

6. Jem Bendell, "Deep Adaptation: A Map for Navigating Climate Tragedy." *Insight Home, University of Cumbria,* July 27, 2018. Revised 2nd Edition released July 27, 2020, https://insight.cumbria.ac.uk/id/eprint/4166/.

In this long and technical paper, Bendell did two unusual things. First, he took some of the more pessimistic climate predictions at face value. Second, he clearly addressed the emotional and existential consequences of those predictions. The most alarming prediction, of huge methane releases from the Arctic, was based on limited observations that have not been replicated. Unfortunately, many other "tipping points", catastrophic and irreversible events that can accelerate climate change, are likely or already under way. The 2020 version of the paper steps back a bit from the dire predictions of the original 2018 version, in response to criticisms.

7. Margi Prideaux, *Fire: A Message From the Edge of Climate Catastrophe* (Mile End, Australia: Wakefield Press, 2022).

Kirk Siegler, "The Camp Fire Destroyed 11,000 Homes. A Year Later Only 11 Have Been Rebuilt," NPR, Nov. 9, 2019, https://www.npr.org/2019/11/09/777801169/the-camp-fire-destroyed-11-000-homes-a-year-later-only-11-have-been-rebuilt. Alice Friedemann, "Delay, Deny, Defend: Why insurance companies don't pay claims." Peak Everything, Overshoot, & Collapse, March 25, 2023, https://energyskeptic.com/2023/delay-deny-defend.

8. Terry Tempest Williams, "What Love Looks Like." Orion, https://orionmagazine.org/article/what-love-looks-like/.

9. Servigne, Pablo and Raphaël Stevens, *How Everything Can Collapse* (Cambridge, UK and Bedford, MA, USA: Polity Press, 2020). First published in French, 2015.

10. "Guy McPherson," RationalWiki, accessed April 2023. https://rationalwiki.org/wiki/Guy_McPherson.

Other predictions cited in this reference:

In 2007 McPherson predicted the USA's trucking industry would collapse by 2012 due to peak oil, quickly followed by the interstate highway system.

In 2008 he predicted the end of civilization by 2018 due to peak oil, "If you're alive in a decade, it will be because you've figured out how to forage locally."

In 2016 he predicted that humanity and most lifeforms would be extinct due to global warming by mid-2026.

11. David Foster Wallace, "This is Water," Kenyon College Commencement Speech, 2005, https://web.ics.purdue.edu/~drkelly/DFWKenyonAddress2005.pdf.

12. David Graeber and David Wengrow, *The Dawn of Everything: A New History of Humanity* (New York: Farrar, Straus and Giroux, 2021).

13. Donella H. Meadows, Dennis L. Meadows, Jorgen Randers, and William W. Behrens III, *The Limits to Growth: A Report for the Club of Rome's Project on the Predicament of Mankind* (Universe Publishing, 1972). The book is available online at: https://donellameadows.org/the-limits-to-growth-now-available-to-read-online.

14. Gaya Herrington, "Update to Limits to Growth: Comparing the World3 Model with Empirical Data." *Journal of Industrial Ecology* 25: 614–626, 2021. https://doi.org/10.1111/jiec.13084. Author's courtesy copy:

https://canadiancor.com/wp-content/uploads/2021/10/Herrington-Limits-2021.pdf.

15. Bayo Akomolafe, Facebook, Sept. 18, 2022.

16. Amitav Ghosh, *The Nutmeg's Curse: Parables for a Planet in Crisis* (Chicago, IL, USA: University of Chicago Press, 2022).

17. Nese O. Ak, Dean O. Cliver, Charles W. Kaspar, "Cutting Boards of Plastic and Wood Contaminated Experimentally with Bacteria," *J. Food Prot.* **57**(1):16-22, 1994. https://doi.org/10.4315/0362-028X-57.1.16.

18. "Tim DeChristopher," Wikipedia, https://en.wikipedia.org/wiki/Tim_DeChristopher.

19. Terry Tempest Williams, "What Love Looks Like." Orion, https://orionmagazine.org/article/what-love-looks-like/. Thanks to Orion magazine for permission to publish this excerpt of the interview of Tim DeChristopher by Terry Tempest Williams.

20. Terry Root has clearly thought long and hard about climate messaging, and I trust that she acted in the way she believed was most responsible. "My Climate Story: Terry Root." Climate One, Sept. 19, 2019, https://www.climateone.org/audio/my-climate-story-terry-root. The intergovernmental and political nature of the IPCC helps it to adopt convenient but obviously erroneous predictions rather than the best predictions for climate change. For a lucid discussion of this, see Sam Hall, "The Busy Worker's Handbook to the Apocalypse," Medium, April 18, 2023, https://medium.com/@samyoureyes/the-busy-workers-handbook-to-the-apocalypse-7790666afde7.

21. Sven De Causmaecker, Svenergy, svenergy.info.

22. T.W. Murphy, D.J. Murphy, T.F. Love, M.L.A. LeHew, and B.J. McCall, "Modernity is incompatible with planetary limits: Developing a PLAN for the future." *Energy Research & Social Science,* 81 (2021). http://doi.org/10.1016/j.erss.2021.102239.

23. *PLAN: Planetary Limits Academic Network*, https://planetarylimits.net/.

24. See, for example, NJ Hagens and DJ White, *Reality Blind: Integrating the Systems Science Underpinning Our Collective Futures,* Volume 1. University of Minnesota (self-published), 2021. Available in bookstores or at: https://read.realityblind.world/view/975731937/i/. Or, with less systems science and more physical science, Thomas W. Murphy, Jr., *Energy and Human Ambitions on a Finite Planet,* UC San Diego Open Educations resources website, 2021. https://escholarship.org/uc/item/9js5291m, https://doi.org/10.21221/S2978-0-578-86717-5.

25. Katie Teague, "Stan Rushworth Interview," In the Making (YouTube), Episode 1, Aug. 18, 2020, https://www.youtube.com/watch?v=Kir_1m-X7OSE&t=1s.

26. "The Great Oxidation Event: How Cyanobacteria Changed Life," *America Society for Microbiology*, Feb. 18, 2022, https://asm.org/Articles/ 2022/February/The-Great-Oxidation-Event-How-Cyanobacteria-Change.

27. The details of past societal collapses and the trajectory of industrial consumer society are active areas of research and debate. The global nature of our predicament makes it different from societal collapses of the past. The USSR's collapse in the 1990's is the basis for: Dmitry Orlov, *The Five Stages of Collapse: Survivors Toolkit* (Gabriola Island, BC, Canada: New Society Publishers, 2013).

28. Dahr Jamail has shared the story of his dying friend as an analogy to the state of the planet in several talks and interviews, including, "The End of Ice and the Climate Crisis: How, Then, Shall We Live." The Real Truth About Health (YouTube channel), June 26, 2020, https://www.youtube.com/ watch?v=akprvVaBu0k.

29. Vanessa Machado de Oliveira, *Hospicing Modernity* (Berkeley, California: North Atlantic Books, 2022).

30. I do not know what comes after death. In hearing a number of near-death experiences of people in my churches, I was more struck by the emotional profundity of those experiences than the variable geography or cast of characters. I suspect that what we will experience is beyond our senses, and that our perception of it is highly variable depending on culture and beliefs.

31. For a number of conditions, hospice care appears to prolong life. See for instance: Stephen R. Connor, Bruce Pyenson, Kathryn Fitch, Carol Spence, Kosuke Iwasaki, "Comparing Hospice and Nonhospice Patient Survival Among Patients Who Die Within a Three-Year Window," *J. Pain and Symptom Management* 33 (3), 238-246 (2007), https://doi.org/10.1016/j.jpainsymman.2006.10.010.

32. This is a brief excerpt from a text Daniela Muhanshim Herzog shared in the Deep Adaptation Facebook group. Muhanshim was diagnosed with leukemia 25 years ago, at age 36. Her son was only seven years old back then. Conventional treatment would have taken her away from her child until the supposed completion of treatment, two years later. Her chances of survival were slim with or without treatment—so she chose not to receive it. Instead, she practiced embracing dying—as belonging naturally to the arc of her life. She tried to create memories for and with her child in the little time she was told she had left. Today she lives and teaches at Furnace Mountain Zen Center and offers "Death Cafés" in the Bluegrass Area of Eastern Kentucky. She can be reached at muhanshim@gmail.com.

33. Adapted from: Connie Barlow, "The Gifts of Death: A Responsive Reading in Celebration of Death," March 2005, The Great Story, https://thegreatstory.org/songs/death-reading-2.html.

34. Ezra Klein, "Your Kids Are Not Doomed," *The New York Times* June 5, 2022, https://www.nytimes.com/2022/06/05/opinion/climate-change-should-you-have-kids.html.

35. The phrase is also the title of several books and of an exhibition in the Smithsonian National Museum of African American History and Culture. "Making a way out of no way," Smithsonian, https://www.si.edu/exhibitions/making-way-out-no-way%3Aevent-exhib-4844 .

36. I offer one classic and one new resource from Black American spirituality for hard times. Thurman, Howard, *Jesus and the Disinherited,* first published in 1949. (Boston, MA, USA: Beacon Press, 1976); Moss III, Otis, *Dancing in the Darkness: Spiritual Lessons for Thriving in Turbulent Times. (*New York, USA: Simon & Schuster, 2023).

37. David Baum, "Community—Jessica Canham," Collapse Club (YouTube channel), Jan. 28, 2022, https://www.youtube.com/watch?v=_CX-Esqj2JHE.

38. The poem, "A Map to the Next World," by U.S. Poet Laureate and citizen of the Muskogee nation, Joy Harjo, uses a native North American tradition that we live in the "fourth world", the others having been destroyed with only a remnant of humanity surviving.
"The Well at the End of the World" and other myths are explored in workshops by Sarah-Jane Menato, https://www.sjmcoachingandtraining.co.uk/workshops/. I will leave the reader to search online for additional stories including "Wetiko/Windigo" and "lifeboat flotilla".

39. This material is adapted from: http://www.aimeemaxwell.net/wp-content/uploads/Everything-Is-Awful-and-I'm-Not-OK-Dr-Aimee-Maxwell.pdf and related materials from Maxwell.

40. Laura Martin, *Breaking into Light* (Washington DC, USA: Politics and Prose, 2023), p. 35. Used by permission.

41. Scott Siskind, "Panic Disorder," Lorien Psychiatry, https://lorienpsych.com/2020/11/29/panic-disorder/.

42. Resmaa Menakem, *My Grandmother's Hands* (Las Vegas, NV, USA: Central Recovery Press, 2017).

43. Therapists and coaches who honor climate-related issues can be found at: Climate Psychology Alliance,
https://www.climatepsychologyalliance.org, and http://guidance.deepadaptation.info.

44. Aimee Maxwell, conversation with David Baum.

45. Did you just yawn after reading that sentence?

46. Robert Karen, *Becoming Attached: First Relationships* (Oxford: Oxford University Press, 1998).

47. See, for example, Diana Divecha, "How Cosleeping Can Help You and Your Baby." Greater Good Magazine, Feb. 7, 2020, https://greatergood.berkeley.edu/article/item/how_cosleeping_can_help_you_and_your_baby.

48. The concept of the non-anxious presence, and the family system dynamics behind it, are key to leadership in anxious times. See: Edwin H. Friedman, *Generation to Generation: Family Process in Church and Synagogue,* (New York: Guilford Press, 1985), especially pp. 208-210. Several authors after Friedman have used this concept.

49. Terry LePage, Grief Gratitude and Courage, https://opendoorcommunication.org/grief.

50. Francis Weller, *The Wild Edge of Sorrow* (Berkeley, CA, USA: North Atlantic Press, 2015).

51. A current book on the Work That Reconnects is: Joanna Macy and Molly Young Brown, *Coming Back to Life: The Updated Guide to the Work That Reconnects.* Gabriola, British Columbia: New Society Publishers, 2014. For Work That Reconnects resources and affiliated events, see https://workthatreconnects.org/.

52. Stephen Jenkinson's work can be found at https://orphanwisdom.com.

53. "Practices," Work that Reconnects, https://workthatreconnects.org/resources/practices/.

54. Mary Nikkel, "In Defense of Frodo: Two Kinds of Strength," Medium, Jan. 11, 2021, https://marynikkel.medium.com/in-defense-of-frodo-two-kinds-of-strength-383d9feec7eb. In this essay, Mary Nikkel examines the character of Frodo in The Lord of the Rings. He has been at close quarters with evil for too long, and it breaks him. Just as we have limits to what we can bear physically, we also have limits mentally, emotionally, and spiritually. Do not carry moral injury alone as Frodo did.

55. Trebbe Johnson, *Fierce Consciousness: Surviving the Sorrows of Earth and Self.* (Ithaca, NY: Calliope Books, 2023), p. 93.

56. Trebbe Johnson, *Radical Joy for Hard Times: Finding Meaning and Making Beauty in Earth's Broken Places.* (Berkeley, CA, USA: North Atlantic Books, 2018).

57. Trebbe Johnson, "Ways to Practice RadJoy." Radical Joy for Hard Times, https://radicaljoy.org/ways-to-practice.

58. Robin Wall Kimmerer, *Braiding Sweetgrass: Indigenous Wisdom, Scientific Knowledge and the Teachings of Plants* (Minneapolis, Minnesota: Milkweed Editions, 2013)

59. A. J. Jacobs, "My journey to thank all the people responsible for my morning coffee." *TED*, https://www.ted.com/talks/a_j_jacobs_my_journey_to_thank_all_the_people_responsible_for_my_morning_coffee?. See also: A. J. Jacobs, *Thanks a Thousand: a Gratitude Journey* (New York: Simon and Schuster/TED, 2018).

60. For the original reference and Jung's explanation of this statement, see: Purrington, "Dr. Jung clarifies misunderstanding of BBC Broadcast of: 'I don't believe. I know,'" *Carl Jung Depth Psychology,* June 3, 2020, https://carljungdepthpsychologysite.blog/2020/06/03/dr-jung-said-i-dont-believe-i-know/.

61. Matthew Fox, *Original Blessing: A Primer in Creation Spirituality Presented in Four Paths, Twenty-Six Themes, and Two Questions.* (Santa Fe, NM, USA: Bear & Company, 1983) and more recent editions.

62. Chogyam Trungpa, *Cutting Through Spiritual Materialism* (Boston, MA, USA: Shambala Publications, 1973.)

63. Wider Embraces, https://widerembraces.org.

64. Kimmerer, *Braiding Sweetgrass.*

65. **Guerrilla gardening** is the act of gardening – raising food, plants, or flowers – on land that the gardeners do not have the legal rights to cultivate, such as abandoned sites, areas that are not being cared for, or private property.

66. Kimmerer, *Braiding Sweetgrass*, p. 180. See also Robin Wall Kimmerer, "The 'Honorable Harvest': Lessons From an Indigenous Tradition of Giving Thanks," Yes! Magazine, Nov. 26, 2015, https://www.yesmagazine.org/issue/good-health/2015/11/26/the-honorable-harvest-lessons-from-an-indigenous-tradition-of-giving-thanks.

67. Jay offers a summary of research findings on the benefits of nature therapy at "Frequently Asked Questions," Held Outside, https://heldoutside.mailchimpsites.com/faq.

68. Laura Martin, *Breaking into Light* (Washington D.C.: Politics and Prose, 2023), p. 41, used by permission.

69. Barbara Cecil and Dahr Jamail, "Transitions: There is Infinite Hope, But Not For Us," Last Born in the Wilderness, #215, Podcast, https://www.lastborninthewilderness.com/episodes/cecil-jamail?fbclid=IwAR2JH5CVA0w4imKce-277JQu_un266isOXQ6UNXKHsn4g-HnowIy4PAiRLs4.

70. Pat McCabe, Woman Stands Shining, is of the Diné Nation. She offers Thriving Life Paradigm as a result of her inquiry into the question, "How do I become that being, that human, whose presence and way of being supports and causes all other life to thrive?" Her recent talks can be found on the web.

71. Participatory Defense, https://www.participatorydefense.org.

72. Jack Zimmerman, Virginia Coyle, *The Way of Council*, Second edition (Spring City, PA, USA: Bramble Books, 2009).

73. Jem Bendell and Katie Carr, "Group Facilitation on Societal Disruption and Collapse: Insights from Deep Adaptation, *Sustainability* 2021, *13*(11), 6280; https://doi.org/10.3390/su13116280.

74. Ways of Council, https://waysofcouncil.net.

75. "Liberating Structures Menu," Liberating Structures. https://www.liberatingstructures.com/ls-menu/.

76. The language of feelings and needs from Nonviolent Communication is very helpful to take blame and shame out of a situation. See Marshall Rosenberg, *Nonviolent Communication: A Language of Life*, 3rd Edition, Encinitas, CA, USA: PuddleDancer Press, 2015.

77. "List of Peace Activists," Wikipedia, https://en.wikipedia.org/wiki/List_of_peace_activists.

78. Alexis Shotwell, *Against Purity: Living Ethically in Compromised Times* (Minneapolis, MN, USA: University of Minnesota Press, 2016).

79. This advice was offered during the online course, "Calling In: Creating Change without Cancel Culture, Loretta J. Ross, https://lorettajross.com/online-courses.

80. "What if Instead of Calling People Out, We Called Then In?" *The New York Times,* November 19, 2020, https://www.nytimes.com/2020/11/19/style/loretta-ross-smith-college-cancel-culture.html.

81. Loretta J. Ross uses the phrase in "Calling in the Calling Out Culture." Her website, http://lorettajross.com, contains online courses, articles and videos about this topic.

82. Shaun Chamberlin, "Confessions of a Hypocrite: Utopia in the Age of Ecocide," Kosmos fall/winter 2016, https://www.kosmosjournal.org/article/confessions-of-a-hypocrite-utopia-in-the-age-of-ecocide/.

83. "Practices," Work that Reconnects, https://workthatreconnects.org/resources/practices/.

84. This quote may be apocryphal: https://quoteinvestigator.com/2013/04/23/good-idea/.

85. Doris Longacre, *Living More with Less*, 30th Anniversary Edition (Scottsdale, PA, USA: Herald Press, 2010).

86. Tyler J. Disney, "One Year of Emergent Renaissance Ecology," TYLERJDISNEY, http://tylerjdisney.com/blog/2021/4/13/one-year-of-emergent-renaissance-ecology.

Fisker's experience and motivations are summarized here: Jacob Lund Fisker, "Early Retirement Extreme: The ten-year update," Get Rich Slowly, https://www.getrichslowly.org/early-retirement-extreme/.

87. Here is Fisker's video explaining his approach. "A Systems Approach to Resilient Lifestyle Design w/ Jacob Lund Fisker." The Stoa (YouTube), Feb. 10, 2021, https://www.youtube.com/watch?v=SPvftqB-WXk.

88. Vicki Robin and Joe Dominguez, *Your Money or your Life* (New York: Penguin Books, first published 1992).

89. Jane Dwinell, *Freedom through Frugality: Spend less, have more* (Addison, VT, USA: Spirit of Life Publishing, 2010).

90. At the time of publication, Jane is still looking for people to join her. Contact me to inquire.

91. Kaitlynn Tiffany, "Nearly all of the big dating apps are now owned by the same company," Vox, Feb. 11, 2019, https://www.vox.com/the-goods/2019/2/11/18220425/hinge-explained-match-group-tinder-dating-apps.

92. Irv Mills, *The Easiest Person to Fool,* http://theeasiestpersonto-fool.blogspot.com.

93. Jessica Canham and David Baum, "Justice, Mutual Aid, & the Living Earth," Collapse Club (YouTube), Jan. 10, 2023, https://www.youtube.com/watch?v=YnDe1AgHOY0.

94. Sven Eberlein, "Bring Transition Town-Style Sharing to your Community," Shareable, January 15, 2013, https://www.shareable.net/bring-transition-town-style-sharing-to-your-community/.

95. Caroline Hickman, Elizabeth Marks, Panu Pihkala, Susan Clayton, R. Eric Lewandowski, Elouise E. Mayall, Britt Wray, Catriona Mellor, Lise van Susteren, *The Lancet: Planetary Health* 5(12), E863-E873, December 2021,https://doi.org/10.1016/S2542-5196(21)00278-3.

96. Britt Wray, *Generation Dread: Finding Purpose in an Age of Climate Crisis* (Toronto, Canada: Knopf Canada, 2022), is good resource. She interviews young adults navigating eco-anxiety and grief.

97. The quote is from a video clip: https://www.instagram.com/reel/Ck-9UOkVvJVN/?igshid=YWJhMjlhZTc%3D.

98. Robert Karen, *Becoming Attached: First Relationships* (Oxford: Oxford University Press, 1998).

99. Alfie Kohn, *The Homework Myth: Why our Kids Get Too Much of a Bad Thing* (Boston, MA, USA: Da Capo Press, 2007).

100. David Willitts, "Have the boomers pinched their children's future?" The Royal Institution (YouTube), Jan. 23, 2020, https://www.youtube.com/watch?v=ZuXzvjBYW8A. See also Omri Wallach, "Charting the Growing Generational Wealth Gap," Dec. 2, 2020, https://www.visualcapitalist.com/charting-the-growing-generational-wealth-gap/.

101. This short story is a cautionary tale for elders. Paolo Bacigalupi, "A Full Life." MIT Technology Review, April 24, 2019, https://www.technologyreview.com/s/613349/a-full-life/.

102. "October Sweet Peas!" Davids Garden Diary, October 21, 2015. https://davidsgardendiary.com/2015/10/21/october-sweet-peas/.

103. To glean means to harvest food left over after the regular harvest, an ancient practice for food justice described in the biblical books of Leviticus 19 and Ruth 2.

104. Terry LePage, CA Native Garden, http://canativegarden.blogspot.com.

105. —, http://canativegarden.blogspot.com/2015/01/indian-burial-mounds.html.

106. —, http://canativegarden.blogspot.com/2015/10/you-must-remember-this-summer-dry.html.

107. —, http://canativegarden.blogspot.com/2015/12/planting-natives.html.

108. —, http://canativegarden.blogspot.com/2015/06/turf-terminators-beware.html.

109. Margi Prideaux, *Fire: A Message From the Edge of Climate Catastrophe* (Mile End, Australia: Wakefield Press, 2022).

110. "A Guide to Growing and Respecting Sacred White Sage." *Flowers by the Sea,* Feb. 7, 2020. https://www.fbts.com/sacred-salvias/guide-to-growing-and-respecting-sacred-white-sage.html.

111. Jean Craighead George, *My Side of the Mountain,* first published 1959 (New York: Puffin Books, 1991).

112. "A food forest, also called a forest garden, is a diverse planting of edible plants that attempts to mimic the ecosystems and patterns found in nature. Food forests are three-dimensional designs, with life extending in all directions – up, down, and out." Definition from Project Food Forest, https://projectfoodforest.org/what-is-a-food-forest/. Food Forest is a concept popular in Permaculture.

113. See, for example, Temple Grandin, "Against Algebra: Students need more exposure to the way everyday things work and are made," *The Atlantic,* October 6, 2022, https://www.theatlantic.com/ideas/archive/2022/10/against-algebra/671643.

114. Joseph Henrich, *The Secret of Our Success: How Culture Is Driving Human Evolution, Domesticating Our Species, and Making Us Smarter* (Princeton, NJ USA: Princeton University Press, 2017).

115. "Seed Freedom: Toward an Earth Democracy, a Conversation with Vandana Shiva," Mold, Nov. 9, 2021, https://thisismold.com/mold-magazine/vandana-shiva-seed-freedom-toward-an-earth-democracy.

116. The issue of digital decay is pervasive and severe, and mostly being ignored. Here are a couple of articles: Teresa Soleau, "Preventing Digital Decay," Getty, Oct. 20, 2014, https://blogs.getty.edu/iris/preventing-

digital-decay/; Patricia Cohen, "Fending Off Digital Decay Bit by Bit," *The New York Times,* March 15, 2010, https://www.nytimes.com/2010/03/16/books/16archive.html.

117. Jared Diamond, *Collapse,* second edition (New York: Penguin Press, 2011).

118. Kirk Siegler, "The Camp Fire Destroyed 11,000 Homes. A Year Later Only 11 Have Been Rebuilt," NPR, Nov. 9, 2019, https://www.npr.org/2019/11/09/777801169/the-camp-fire-destroyed-11-000-homes-a-year-later-only-11-have-been-rebuilt.

119. Alice Friedemann, "Delay, Deny, Defend: Why insurance companies don't pay claims." Peak Everything, Overshoot, & Collapse, March 25, 2023, https://energyskeptic.com/2023/delay-deny-defend.

120. For example, fire insurance in California: Yanjun (Penny) Liao, Margaret A. Walls, Matthew Wibbenmeyer, and Sophie Pesek, "Insurance Availability and Affordability under Increasing Wildfire Risk in California." Resources for the Future, Nov. 30, 2022, https://www.rff.org/publications/issue-briefs/insurance-availability-and-affordability-under-increasing-wildfire-risk-in-california/.

121. For a background on Eco Art, see this group of of several hundred artists, art historians, scientists and writers that has been growing since its establishment in the 1990s: https://www.ecoartnetwork.org/. See also: *EcoArt in Action,* Geffen, Rosenthal, Fremantle and Rahmani, eds. (New York: New Village Press, 2022). See also Women's EcoArt Dialogue (WEAD), https://www.weadartists.org/ and ecoartspace, which is a membership organization for any gender that includes artists, art historians, scientists, and writers https://ecoartspace.org/. Both do online and in gallery exhibitions, publish books and catalogs and offer many services for members.

122. Molly Housh Gordon, "How to Survive the End of the World: A Climate Strike Sermon." Medium, https://mollyhoushgordon.medium.com/how-to-survive-the-end-of-the-world-61f5119fcce2.

123. Maxine Hong Kingston, *The Fifth Book of Peace* (New York: Vintage International, 2004.) The book was written in response to her home, including a manuscript of a novel called *The Fourth Book of Peace,* being lost in a fire: https://www.spiritualityandpractice.com/books/reviews/view/6803?id=6803.

ACKNOWLEDGMENTS

I have received support from many people, without whom this book could not have been written. First, Jem Bendell for the framework of Deep Adaptation and the founding of the Forum, April Tuck and the wise ones who gave me this book to write, Thea Gavin for leading her Grand Canyon Writer's Retreat that got this project moving, and Sharon Graff for coaching me to create Grief Gratitude and Courage. I am also grateful for the authors, speakers, bloggers, and podcasters who have been making sense of our predicament, only some of whom are mentioned in the bibliography.

I am so grateful to my many facilitation mentors and collaborators past and present, especially Emma Mary Gathergood, Kat Soares, Katie Carr, Lisa Gordon, Margo Finlayson, and Sophie Reynolds. And to the DA Resilience Accountability Group, especially Dan, Daniel, Jane, Mary Jo, and Tom.

To Lynne Prechel who kept me centered in Spirit, Yoojin Lee who prayed and empathized with me, Kim Weiss who encouraged me and was a champion editor, Jane Dwinell who supported me over the yearlong project and advocated for lists, Vivian Johnson who helped me sort out the introduction, Lisa Gordon who provided wise editing, and to all those who read the manuscript and offered suggestions.

To all the authors who allowed me to include their published or unpublished work: Aimee Maxwell, Connie Barlow, Daniela Muhanshim Herzog, Ellen Rosner, Irv Mills, Jenna Matlin, Karen Perry, Laura Martin, LZ, Margi Prideaux, Michael Dowd, Tom Schloegel, the estate of Carol Bialock, and the kind folks at Orion Magazine. Special thanks to those whose answers to questions on the Deep Adaptation Facebook page I included or excerpted as "many voices".

To all who allowed me to interview them, and to publish excerpts of those interviews: Andrew Martelle, Ari Jong, Brennan Smith, Carla Brennan, Daniel Kim, Ellen Wilson, Emma Mary Gathergood, Gwen Fischer, Lucas Fischer, Jane Dwinell, Jay Ridgewell, Jessica Canham,

Jonathan Kabat, Josie Bennett, Kat Soares, Lisa Gordon, Lucas Fischer, Margi Prideaux, Peter Kindfield, Raewyn Proctor, Saber Freeman, Sven De Causmaecker, and Wendy Freeman.

I am grateful to the participants of circles and workshops who have shared their hearts with me.

And I am most grateful to my partner of many years Scott Rychnovsky, who always supports and encourages me, and has been telling me for years that I should write a book.

BEYOND THIS BOOK

You are welcome to contact the author at terry@opendoorcommunication.org. She may be able to match you with resources appropriate to your needs and interests. Some of her groups can be found at deepadaptation.info/events; you are invited to join.

Opendoorcommunication.org/eye is the website for this book. On this website you will find:

- Ordering options for this book in paperback and electronic forms, including international orders,

- A copy of the endnotes with clickable links to websites, and

- Additional resources related to this book and some of the topics in it.

Printed in the USA
CPSIA information can be obtained
at www.ICGtesting.com
LVHW011141080324
773848LV00004B/950